Axiomatic Consensus Theory in Group Choice and Biomathematics

FRONTIERS
IN APPLIED MATHEMATICS

The SIAM series on Frontiers in Applied Mathematics publishes monographs dealing with creative work in a substantive field involving applied mathematics or scientific computation. All works focus on emerging or rapidly developing research areas that report on new techniques to solve mainstream problems in science or engineering.

The goal of the series is to promote, through short, inexpensive, expertly written monographs, cutting edge research poised to have a substantial impact on the solutions of problems that advance science and technology. The volumes encompass a broad spectrum of topics important to the applied mathematical areas of education, government, and industry.

BOOKS PUBLISHED IN FRONTIERS IN APPLIED MATHEMATICS

Day, William H. E., and McMorris, F. R., *Axiomatic Consensus Theory in Group Choice and Biomathematics*

Banks, H.T. and Castillo-Chavez, Carlos, editors, *Bioterrorism: Mathematical Modeling Applications in Homeland Security*

Smith, Ralph C. and Demetriou, Michael, editors, *Research Directions in Distributed Parameter Systems*

Höllig, Klaus, *Finite Element Methods with B-Splines*

Stanley, Lisa G. and Stewart, Dawn L., *Design Sensitivity Analysis: Computational Issues of Sensitivity Equation Methods*

Vogel, Curtis R., *Computational Methods for Inverse Problems*

Lewis, F. L.; Campos, J.; and Selmic, R., *Neuro-Fuzzy Control of Industrial Systems with Actuator Nonlinearities*

Bao, Gang; Cowsar, Lawrence; and Masters, Wen, editors, *Mathematical Modeling in Optical Science*

Banks, H. T.; Buksas, M. W.; and Lin, T., *Electromagnetic Material Interrogation Using Conductive Interfaces and Acoustic Wavefronts*

Oostveen, Job, *Strongly Stabilizable Distributed Parameter Systems*

Griewank, Andreas, *Evaluating Derivatives: Principles and Techniques of Algorithmic Differentiation*

Kelley, C. T., *Iterative Methods for Optimization*

Greenbaum, Anne, *Iterative Methods for Solving Linear Systems*

Kelley, C. T., *Iterative Methods for Linear and Nonlinear Equations*

Bank, Randolph E., *PLTMG: A Software Package for Solving Elliptic Partial Differential Equations. Users' Guide 7.0*

Moré, Jorge J. and Wright, Stephen J., *Optimization Software Guide*

Rüde, Ulrich, *Mathematical and Computational Techniques for Multilevel Adaptive Methods*

Cook, L. Pamela, *Transonic Aerodynamics: Problems in Asymptotic Theory*

Banks, H. T., *Control and Estimation in Distributed Parameter Systems*

Van Loan, Charles, *Computational Frameworks for the Fast Fourier Transform*

Van Huffel, Sabine and Vandewalle, Joos, *The Total Least Squares Problem: Computational Aspects and Analysis*

Castillo, José E., *Mathematical Aspects of Numerical Grid Generation*

Bank, R. E., *PLTMG: A Software Package for Solving Elliptic Partial Differential Equations. Users' Guide 6.0*

McCormick, Stephen F., *Multilevel Adaptive Methods for Partial Differential Equations*

Grossman, Robert, *Symbolic Computation: Applications to Scientific Computing*

Coleman, Thomas F. and Van Loan, Charles, *Handbook for Matrix Computations*

McCormick, Stephen F., *Multigrid Methods*

Buckmaster, John D., *The Mathematics of Combustion*

Ewing, Richard E., *The Mathematics of Reservoir Simulation*

Axiomatic Consensus Theory in Group Choice and Biomathematics

William H. E. Day
Port Maitland, Nova Scotia, Canada

F. R. McMorris
Illinois Institute of Technology
Chicago, Illinois

siam
Society for Industrial and Applied Mathematics
Philadelphia

10 9 8 7 6 5 4 3 2 1

Library of Congress Cataloging-in-Publication Data

Day, William H. E.

Axiomatic consensus theory in group choice and biomathematics / William H.E. Day, F.R. McMorris.

p. cm. — (Frontiers in applied mathematics ; 29)

Includes bibliographical references (p.).

ISBN 0-89871-551-2 (pbk.)

1. Biomathematics. 2. Axiomatic set theory. 3. Consensus (Social sciences) 4. Social choice. I. McMorris, F. R. II. Title. III. Series.

QH323.5 .D39 2003
570'.15'1—dc22

2003061618

Contents

List of Figures

List of Tables

Foreword

I feel more than a little responsible for the genesis of this book. The term "bioconsensus" was invented by Fred S. Roberts, the Director of DIMACS. He and I helped to organize two workshops on bioconsensus at DIMACS, and this book evolved from a tutorial given at one of them. Apart from that, some fifteen years ago I conjectured that it ought to be possible to construct a mathematical framework in which many of the known versions of Arrow's impossibility theorem could be established. That turned out to be a difficult and subtle project on which several of us have labored through the years. Progress has been made; much remains to be done.

Since consensus theory has its roots in the theory of elections, many contributions have been (and are being) made by political scientists, sociologists and economists. In the context of human behavior, consensus research is carried out by psychologists. Faced with conflicting evidence on evolutionary history, systematic biologists appeal to concepts of consensus, and molecular biologists attempt to apply consensus theory in areas of DNA research. Market researchers find the discipline relevant since it can be viewed as a theory of how large organizations make decisions based on possibly conflicting lines of evidence. Contemporary applications of consensus theory involve airplane and missile navigational systems, methods to combat bioterrorism, drug development based on DNA research, marketing and manufacturing decisions of large companies, and stock market predictions. There is a demonstrated need for research in consensus theory, and indeed it is an active and lively area of endeavor in several disciplines.

But the situation is unusual because we have an essentially mathematical framework that was developed by people who were not formally trained as mathematicians. The literary style in the social and behavioral sciences can be quite different from the formalism that some mathematicians have grown to expect. Definitions are sometimes embedded in discussions and not formally delineated; symbolism is not often what one expects, and implied parentheses may be ambiguous. Nevertheless, when one reads papers in different disciplines, one is struck by the similarity of the arguments. In consensus theory there is a pressing need for mathematicians to develop appropriate models in which all of the relevant concepts can be embedded, compared, and analyzed.

I liken the situation to that of measure theory before it was developed as a mathematical discipline. The rudiments were there in the classical theory of real variables, but the subject was difficult because people used subtle results from the reals when they did not really need them. The Lebesgue integral became easier to understand in terms of a measure algebra than it ever was as a branch of classical real analysis. What was a second-year graduate course could suddenly be treated at the advanced undergraduate level. Galois theory as a tool for understanding solvability of polynomials is another example. Indeed, twentieth-century mathematics is distinguished by the axiomatic approach. It is replete with examples where abstract axiom systems make it easier to understand what is happening. Sometimes one uses unneeded machinery when a subject is embedded in a concrete framework; certainly this has been true for consensus theory.

The mathematics discussed by Day and McMorris is sound and has a unique flavor. Often positive results are attained when something that looks unpleasant happens; often one encounters a situation where if X and Y happen, then some apparently desirable thing Z cannot happen, or some undesirable thing W must necessarily occur. A collection of sets that is closed under unions and intersections may be undesirable, while closure under intersection but not union may turn out to be what is needed. For those of us trained in algebra this gives the subject a distinctive flavor. In group theory we are largely interested in types of subgroups, rather than subsets that are not subgroups. In consensus theory we get mileage out of ternary relations having the property that abc, bca, and cab cannot simultaneously occur. The subject is in its infancy. Almost everywhere one looks there are open questions. For example, consensus methods often operate on collections of subsets of a fixed set. It has been useful to represent these collections in terms of ternary relations, to form the consensus in terms of these relations, and then to recapture the consensus object. Yet there is no axiomatic theory of these relational representations, and until very recently no one seems to have worried about just when such representations were injective.

Consensus theory involves good mathematics that is exciting, worthwhile, and potentially applicable to other areas. Yet our current course structure does not expose our students to it, and since the literature is so multidisciplinary, it is not so easy to learn on one's own. And there is a certain amount of intellectual snobbery that makes some mathematicians uncomfortable with the classical consensus literature. All of this cries out for a modern axiomatic treatment that will make things easier and more transparent to the mathematical community. This is precisely what Day and McMorris have undertaken to present. I know them well: they have a deep knowledge of the subject matter, are careful workers, and are good expositors. This book describes the frontier of a new and important branch of applied mathematics, and with its extensive bibliography it provides an indispensable guide to the consensus literature.

I feel a sense of wonderment when I see the parallels between results in voting theory and corresponding results for unrooted and rooted trees. There is something fundamental driving all of this, and Day and McMorris's book helps to reveal these fundamental concepts. As such, their book will be an invaluable resource for researchers in bioinformatics, for evolutionary biologists, for ecologists, for students of discrete mathematics, for computer scientists, and for workers in fields related to social welfare economics.

M. F. Janowitz, Associate Director
Center for Discrete Mathematics and Theoretical Computer Science

Preface

I will briefly tell you about the principles of the art of election.
— Heroine of a 13th century romance by Ramon Lull [257, p. 100]

I will leave out much, but nothing of significance. — I. Pears [325, p. 3]

Ubiquitous in data analysis are problems of aggregation such as this one: Given a set $\mathcal{X}$ of objects of interest, identify an appropriate consensus rule $C : \mathcal{X}^k \longrightarrow \mathcal{X}$ that, when presented a k-tuple P of objects of $\mathcal{X}$, returns from $\mathcal{X}$ a unique consensus object that in some sense best represents P. Solving this problem requires that we evaluate consensus rules in terms of the basic properties or axioms that they satisfy. But for this problem one can sometimes obtain a fundamental impossibility result by first listing a set of seemingly reasonable axioms that any consensus rule on $\mathcal{X}$ should satisfy, then proving that no such rule can exist. When it occurs, this contradiction encourages researchers to explore the extent to which the axioms can be weakened, while still maintaining the contradiction, and to explore how to alter the axioms so as to eliminate the contradiction. Desirable consequences of the latter analysis are characterizations of consensus rules in terms of axioms with which users may assess the appropriateness of the rules for given applications. Thus, although the axiomatic approach may seem to be abstract and purely technical, its practical and concrete aspects enable us to distinguish between what is realizable and what is not.

An archetypal problem of aggregation concerns group choice, i.e., the reduction of different individual preferences to a single collective or group preference. In 1950 K. J. Arrow pioneered the use of axiomatic paradigms in group choice where, for $\mathcal{X}$ a set of preference rankings, he established that there cannot exist a rational efficient democratic rule to obtain a consensus of such rankings. His paradigms have since been refined, extended, and applied in areas far removed from their origins in group choice and welfare economics. We review this axiomatic research first in the area of group choice, then in areas of biomathematics where the objects of interest represent, e.g., partitions of a set, hierarchical structures, unrooted trees, or molecular sequences. We establish the axiomatic consensus theory's basic results using standard terminology and notation and with uniform attention to rigor and detail. We cite both traditional and current literature, and we pose open problems that remain to be solved.

Through the years we have benefited immeasurably from consensus-related discussions and collaborations with J. P. Barthélemy, M. F. Janowitz, B. Leclerc, B. G. Mirkin, B. Monjardet, H. M. Mulder, R. C. Powers, and F. S. Roberts. We view this book both as a culmination of these many beneficial interactions and as a consequence of our tutorial [153] on 2 October 2001 in a meeting on bioconsensus [209] at the Center for Discrete Mathematics and Theoretical Computer Science (DIMACS), Rutgers University. The book's preparation was supported in part by DIMACS and by the Office of Naval Research (grant N00014-00-1-0004). We are particularly indebted to M. F. Janowitz, B. Leclerc, and R. C. Powers for their valuable critiques of the book in a preliminary version.

William H. E. Day
F. R. McMorris

Chapter 1
Achieving Consensus

In all these fields the axiomatic approach—inspired by Arrow's work—is the same. One chooses axioms that the consensus function should satisfy and one determines the functions (if any) satisfying the axioms. — B. Monjardet [296, p. 52]

However, consensus methods used in molecular biology tend to be chosen because they seem interesting or useful, rather than on the basis of some theory.
— B. G. Mirkin and F. S. Roberts [287, p. 711]

Within the broad purview of biomathematics, where mathematical methods are applied to biology and medicine, this book focuses on bioconsensus, a rapidly evolving field in which consensus methods, having been developed for use in social choice theory, are adapted for use in areas such as systematic biology and evolutionary or molecular biology. We will review the history and development of this field, emphasizing how axiomatic consensus theory has flourished in, and contributes to our understanding of, group choice and bioconsensus. We set the stage with typical examples.

Example 1.1. A benevolent society wishes to adopt a rule to achieve consensus so that, when choosing from among various alternatives, the society's officials can apply the rule to members' rankings of the alternatives and so obtain a social ranking that reveals the society's collective will. After discussion the members identify properties of collective rationality, independence, optimality, and democracy that the rule should satisfy.

1. The consensus rule should return a transitive social ranking for all tuples of members' rankings of the alternatives.
2. When voting is imminent, some alternatives may be no longer available. If each member has ranked the available alternatives the same in one situation as in another, then no matter what is true about the rankings of unavailable alternatives, the social ranking of the available alternatives should be the same in both situations.

3. If every member prefers alternative x to alternative y, then the social ranking should prefer x to y.
4. The social ranking should not always be the ranking of a dictatorial member.

The committee evaluating rules for possible adoption learns that no consensus rule can satisfy these seemingly reasonable properties. To fathom why they are contradictory, the committee considers a natural way to achieve consensus by means of the majority rule: for each pair (x, y) of alternatives, x should be socially preferred to y if a majority of members would choose x over y if only those alternatives were available. But the social ranking so derived can violate the transitivity required by property 1 on the preceding page, so it does not exhibit the collective rationality required of an adopted consensus rule. ■

Although Example 1.1 is apocryphal, it encourages us to investigate the properties of proposed consensus rules, and it cautions us that sets of desirable properties need not be logically consistent. Furthermore the problems exhibited by the example arise not just in group choice but in all fields where aggregation problems arise. Suppose that each individual is an expert systematist, each preference is a hierarchical classification, and the social choice is a consensus classification: such problems are relevant in fields of data analysis, classification, and systematics. Consensus problems occur as well in bioinformatics or computational biology [151].

Example 1.2. Waterman et al. [411, 171, 408] describe a method to find unknown consensus patterns occurring imperfectly in both composition and location in a set of molecular sequences. The sequences are approximately aligned and contain unknown patterns that occur (perhaps imperfectly) in each sequence. Waterman's algorithm moves a window of fixed width along the aligned sequences. At each window position it searches for short candidate consensus patterns that, within the window, minimize the total number of mismatches to substrings of each sequence. From among the candidate consensus patterns identified at all window positions, the algorithm selects those consensus patterns with the minimum total number of mismatches.

Since the algorithm is parameterized by the number of symbols from which the sequences are formed, the number of sequences, the length of the aligned set of sequences, the window width, the consensus pattern length, and the number of mismatches permitted when comparing consensus pattern to sequence substring, it is hard to visualize how the algorithm compares with those solving other consensus problems. To understand Waterman's method, Day and McMorris [152] formulated it as a consensus substring (CST) decision problem and compared it with the well-known longest common substring (LCST) problem [174, p. 228]. Waterman's CST algorithm can be seen to generalize an efficient algorithm solving the LCST problem. Mirkin and Roberts [287] show that well-known consensus rules of the social sciences, based on the median and mean, are special cases of Waterman's method for appropriate choices of its parameter values and that Waterman's recommended parameter values cause his method to be based on the median consensus rule. It might be useful to axiomatize Waterman's method or the median rule in the context of his method [287, p. 711]. ■

Example 1.3. Bains [25, 26] describes a heuristic algorithm to align sets of either DNA or protein sequences. The algorithm iterates between an alignment step, which generates a new sequence alignment using the current consensus sequence, and a consensus step, which generates a new consensus sequence using the current sequence alignment. These steps are unapologetically heuristic, e.g., the consensus step in [25] applies seven heuristically developed rules to generate a new consensus base at each aligned position. To understand better the behavior of such algorithms, one might investigate the abstract properties required by each step to reveal the algorithm's theoretical bases, to ensure its convergence, and to achieve for it an acceptable computational complexity. ■

This book stresses formal aspects of the problem of achieving consensus. We introduce the basic concepts, survey the literature, and suggest topics for further study; but for consensus methodology we describe neither its many applications in the biological sciences nor its many associated advances in algorithm design and computational complexity. We assume that readers have a certain familiarity with mathematical reasoning, a certain capacity for abstract thought, and a certain interest in how mathematics is used to model aspects of biological problems. We state many theorems, cite authors and sources, and prove the main results.

Theoretical investigations of problems in group choice are not new: still cited today are contributions of Borda [100] in 1784, Condorcet [137] in 1785, and Pareto [321] in 1896. But the recent origins of our interest in consensus lie in K. J. Arrow's influential doctoral thesis [11], which appeared in 1951 and launched him on a path to share with J. R. Hicks the 1972 Nobel Memorial Prize in Economic Science. Arrow's brilliant insight was to visualize the preference aggregation problem axiomatically so as to prove that there is no rational, efficient, democratic rule of preference aggregation. We will study such abstract formulations of consensus. In this chapter we describe formal models of consensus rules and introduce axiomatic paradigms with which to investigate them. In Chapter 2 we use the paradigms to investigate Arrow's impossibility theorem for weak orders, which generalize the rankings in Example 1.1. In Chapter 3 we apply Arrow's paradigms to obtain impossibility results for basic problems of biological consensus. In Chapter 4 we apply Arrow's paradigms to obtain possibility results, i.e., characterizations, for well-known bioconsensus rules. In Chapter 5 we present formalizations of Arrow's paradigms that reveal general conditions under which the paradigms can be applied. Such studies reveal a shift of perspective: where previously our investigation of an object's internal structure may have yielded insights into the relationships among those objects, now we use object interrelationships to gain insights into the basic properties of sets of those objects. In Chapter 6 we extend the consensus concept to problems where consensus objects may include subsets or supersets of the alternatives associated with the individual objects; such problems arise in systematic biology [9, 364, 363] and may benefit from additional axiomatic investigations of possibilities and impossibilities.

1.1 Consensus Rules

Because consensus problems arise in disparate fields of the social, biological, and management sciences, readers will encounter in the literature a plethora of names and notations for common concepts. Although we may mention variants, we will use a standard terminology and notation beginning with the basic concept of consensus. Let $\mathcal{X}$ be a generic set of objects to be aggregated, let $K = \{1, \dots, k\}$ be a set of *individuals*, and let $\mathcal{X}^k$ be the set of all

k-tuples of $\mathcal{X}$. The tuple $(x_1, \ldots, x_k) \in \mathcal{X}^k$ is called a *profile*, in which object x_i corresponds to individual i, and $(x)_k$ is the *constant* profile $(x_1, \ldots, x_k)$, where $x_1 = x_2 = \cdots = x_k = x$.

Convention 1.4. *Unless specifically stated otherwise, the set K of individuals is finite with $|K| = k \geq 2$.*

In the early chapters we consider consensus rules

$$C : \mathcal{X}^k \longrightarrow \mathcal{X},$$

which map a profile of k individual objects to a single consensus object. Later we study consensus rules that may accept profiles of varying lengths or may return more than one object as a consensus result. To adapt to profiles of varying lengths let $\mathcal{X}^*$ be the set of all profiles of finite positive length, i.e., $\mathcal{X}^* = \bigcup_{k \geq 1} \mathcal{X}^k$, and consider consensus rules

$$C : \mathcal{X}^* \longrightarrow \mathcal{X}.$$

To distinguish between the domains $\mathcal{X}^k$ and $\mathcal{X}^*$, we call a consensus rule with domain $\mathcal{X}^*$ a *complete* consensus rule. To adapt to cases where a consensus result need not be unique, let $2^{\mathcal{X}} \setminus \{\emptyset\}$ be the set of all nonempty subsets of $\mathcal{X}$ and consider consensus rules

$$C : \mathcal{X}^k \longrightarrow 2^{\mathcal{X}} \setminus \{\emptyset\}.$$

To distinguish between the codomains $\mathcal{X}$ and $2^{\mathcal{X}} \setminus \{\emptyset\}$, we call a consensus rule with codomain $2^{\mathcal{X}} \setminus \{\emptyset\}$ a *multiconsensus* rule. Beginning in section 4.3 we consider complete multiconsensus rules

$$C : \mathcal{X}^* \longrightarrow 2^{\mathcal{X}} \setminus \{\emptyset\},$$

which map a profile of individual objects to a nonempty set of consensus objects.

We view the generic elements of $\mathcal{X}$ as atomic indecomposable objects that are distinguishable from each other but otherwise structureless and featureless. However, in specific cases the objects of interest may be quite complex; typically we specify a finite nonempty set S of alternatives (outcomes, bases, evolutionary units, sites, species, vertices) on which the objects are based. For example, objects may be types of n-ary relations (Chapter 2, sections 3.1–3.3), hierarchical structures based on sets of subsets (sections 3.4–3.5, Chapter 4), order-theoretic structures based on partial orders (Chapter 5), or graph-theoretic structures based on concepts of proximity (section 5.3). We define such object types as the need arises, but since Chapters 2 and 3 employ n-ary relations, here we give their common setting.

Convention 1.5. *If $X = \{a, b, \ldots, y, z\}$ is any set of elements denoted by single letters (or digits), we may shorten its representation to $X = ab \cdots yz$. For example, $Y = \{\{a, c, d\}, \{a, f\}, \{b, d, f, g\}, \{e\}\} = \{acd, af, bdfg, e\}$. Also we may shorten any ordered pair (a, b) to ab. Context will determine whether ab means $\{a, b\}$ or (a, b).*

Definition 1.6. *Let $S = S_n$ be a finite set with $n = |S| > 0$, where $|S|$ denotes the cardinality of S.* A binary relation on *S is a subset $R \subseteq S^2$ of ordered pairs. To show membership in R we write $(a, b) \in R$, $ab \in R$, or aRb.*

Table 1.1. *Logical Notation. P and Q are logical sentences; v is a variable.*

Usage	Meaning
$\neg P$	It is not the case that P.
$P \wedge Q$	P and Q.
$P \vee Q$	P or Q, i.e., $P \vee Q$ is true unless P and Q are each false.
$P \implies Q$	P logically implies Q.
$P \iff Q$	P and Q are logically equivalent.
$(\forall v) P$	For every v, P.
$(\exists v) P$	For some v, P.
[...] (...)	Paired delimiters used to prevent ambiguity of meaning.

To investigate binary relations we use elementary logical notation (Table 1.1) and interpret complex logical sentences by means of the following convention.

Convention 1.7. *To reduce the paired parentheses required to specify the meaning of logical sentences, we apply logical operators in the order first $\neg$, then $\vee$ and $\wedge$, finally $\implies$ and $\iff$; thus $\neg P \wedge Q$ means $(\neg P) \wedge Q$ rather than $\neg(P \wedge Q)$. If $P = xRy$ for a binary relation R on S, then of course $\neg P = \neg xRy = \neg(xRy)$. Parentheses always delimit the formula to which quantifiers apply; thus in $(\exists x)(x < y) \vee y = 0$ the quantifier applies only to $(x < y)$.*

First we define basic properties of binary relations (Table 1.2); then we specify several types of relations that occur in problems of discrete applied mathematics (Table 1.3). We illustrate these properties and types with binary relations on a binary set (Table 1.4), but there will be more detailed discussions of weak orders (in Chapter 2), equivalence relations (in section 3.1), and tree quasi-orders (in section 3.2).

1.2 Paradigms

Several paradigms exist to investigate consensus problems. One might formulate a consensus rule to exhibit some desirable features, then analyze that rule to identify other strengths or weaknesses; or one might formulate a set of axioms or properties that many researchers would accept as desirable, then determine the set of consensus rules satisfying those axioms. Arrow used the latter paradigm and for the most part so will we, but its application is sensitive to the questions being asked and to the relative strengths of the axioms involved.

Formulating a viable set of axioms is something of an art. Alone each axiom should be compelling. If the set is too strong, no consensus rule can satisfy all the axioms; if the set is too weak, the set of consensus rules satisfying the axioms may be too large or too unorganized to be useful. In Arrow's analysis, axioms of independence and optimality are separately compelling but, taken together, are so strong that they yield an unsatisfactory set of dictatorial consensus rules. The result can be expressed as in the following templates.

Table 1.2. *Properties of Binary Relations R on S*

Property	Definition
Reflexivity	$(\forall x \in S)(xRx)$
Irreflexivity	$(\forall x \in S)(\neg xRx)$
Completeness	$(\forall x, y \in S)(xRy \vee yRx)$
Symmetry	$(\forall x, y \in S)(xRy \Longrightarrow yRx)$
Antisymmetry	$(\forall x, y \in S)(xRy \wedge yRx \Longrightarrow x = y)$
Transitivity	$(\forall x, y, z \in S)(xRy \wedge yRz \Longrightarrow xRz)$
Tree condition	$(\forall x, y, z \in S)(xRy \wedge xRz \Longrightarrow yRz \vee zRy)$

Table 1.3. *Types of Binary Relations R on S_n. Also given is a symbol for the set of all relations on S_n of this type.*

Type	Defined as a ...	Symbol
Binary	Subset $R \subseteq S_n^2$	$\mathcal{R} = \mathcal{R}_n$
Equivalence	Reflexive, symmetric, transitive binary relation	$\mathcal{E} = \mathcal{E}_n$
Weak order	Complete, transitive binary relation	$\mathcal{O} = \mathcal{O}_n$
Partial order	Reflexive, antisymmetric transitive binary relation	$\wp = \wp_n$
Quasi-order	Reflexive, transitive binary relation	$\mathcal{Q} = \mathcal{Q}_n$
Tree quasi-order	Quasi-order satisfying the tree condition	$\mathcal{T} = \mathcal{T}_n$

Table 1.4. *Binary Relations R_i on $S_2 = \{a, b\}$. For the types of relations defined in Table* 1.3, $\mathcal{E}_2 = \{R_7, R_{15}\}$, $\mathcal{O}_2 = \{R_{12}, R_{13}, R_{15}\}$, $\wp_2 = \{R_7, R_{12}, R_{13}\}$, *and* $\mathcal{Q}_2 = \mathcal{T}_2 = \{R_7, R_{12}, R_{13}, R_{15}\}$.

i	R_i	Refl.	Compl.	Sym.	Antisym.	Trans.	Tree cond.
0	$\emptyset$			√	√	√	√
1	$\{aa\}$			√	√	√	√
2	$\{ab\}$				√	√	
3	$\{ba\}$				√	√	
4	$\{bb\}$			√	√	√	√
5	$\{aa, ab\}$				√	√	
6	$\{aa, ba\}$				√	√	√
7	$\{aa, bb\}$	√		√	√	√	√
8	$\{ab, ba\}$			√			
9	$\{ab, bb\}$				√	√	√
10	$\{ba, bb\}$				√	√	
11	$\{aa, ab, ba\}$			√			
12	$\{aa, ab, bb\}$	√	√		√	√	√
13	$\{aa, ba, bb\}$	√	√		√	√	√
14	$\{ab, ba, bb\}$			√			
15	$\{aa, ab, ba, bb\}$	√	√	√		√	√

Template 1.8. *If a consensus rule has the desirable properties X and Y, then it also has the undesirable property* $\neg Z$.

Template 1.9. *No consensus rule can have the desirable properties X, Y, and Z.*

The impossibility results we describe include Theorems 2.9, 2.15, 2.18, 3.8, 3.20, 3.32, 3.52, 3.66, 4.32, 6.7, 6.8, and 6.9. But how might desirable rules (if they exist) be characterized? We might sequentially introduce relatively weak axioms of symmetry, neutrality, monotonicity, or the like until they yield a meaningful set of consensus rules. The result can be expressed as in the following theorem.

Template 1.10. *Consensus rule C is the unique rule* $\widehat{C}$ *if and only if C has the desirable properties X, Y, and Z.*

Characterizing a set of many rules may be valuable. For example, the majority and strict consensus rules are often used to take the consensus of profiles of hierarchies. There exist parameterized sets of consensus rules for hierarchies that include the majority and strict rules as extremes. Characterizing the parameterized set might yield useful characterizations of its extremes. The possibility results we describe include characterizations of unique rules (Theorems 2.24, 4.23, 4.51, 4.52, 5.31, 5.35, 5.44, and 5.45; Corollaries 3.9 and 4.41) and of sets of rules (Theorems 4.8, 4.9, 4.17, 4.40, 5.17, and 5.21; Corollaries 5.18 and 5.22). But in a twilight zone between impossibility and possibility are ambiguous results as in the following template.

Template 1.11. *A consensus rule has the desirable properties X, Y, and Z if and only if it also has the undesirable properties V and W.*

Such results include Theorems 3.8, 3.37, and 3.58 and Corollaries 3.38 and 3.70.

1.3 Axioms

Since axioms are essential, we take great care in their specification. We use logical notation (Table 1.1 on page 5) to achieve precision and brevity in axiomatic specifications and in proofs. Our approach is pragmatic and informal: we use logical symbols if the result is easier to understand than the (longer or less clear) formulation without symbols. The axioms are defined in tables that concern particular settings (Table 1.5). An index (Table A.3 on page 116) gives the abbreviation of each axiom and the pages on which the axiom is defined. Regrettably there has been little effort to develop, and less success in achieving, a standard nomenclature for consensus axioms; when browsing the consensus literature, the reader must be prepared to find axiomatic concepts disguised by a variety of (sometimes perplexing) aliases. For example, our axioms of decisive neutrality, neutrality, and S-neutrality are also called [141] neutrality, profile stability, and permutation compatibility, respectively.

Table 1.5. *Axioms: Settings*

Table	Object of Interest	Focus
2.3 on p. 14	Weak order	Impossibility results for consensus rules
3.1 on p. 29	Equivalence relation	" " "
3.2 on p. 35	Tree quasi-order	" " "
3.3 on p. 39	Unrooted phylogeny	" " "
3.4 on p. 45	Hierarchy	" " "
4.1 on p. 54	Hierarchy	Counting consensus rules
4.2 on p. 62	Hierarchy	Intersection consensus rules
4.3 on p. 73	Hierarchy	Median complete multiconsensus rules
5.2 on p. 86	Meet semilattice	Projection and federation consensus rules
5.4 on p. 92	Median graph	Median complete multiconsensus rules
5.5 on p. 97	Meet semilattice	" " "
6.1 on p. 108	Unrooted phylogeny	Impossibility results for generalized rules

1.4 Naming and Finding

Chapters are named by positive integers, e.g., 1, 2, 3; a chapter's sections, by a subordinate integer, e.g., 2.1, 2.2, 2.3; a section's subsections, by a second subordinate integer, e.g., 2.1.1, 2.1.2, 2.1.3. Within each chapter, one series of integers names that chapter's tables, e.g., Tables 3.1, 3.2, 3.3 in Chapter 3; a second names that chapter's figures, e.g., Figures 3.1, 3.2, 3.3 in Chapter 3; a third names the conventions, corollaries, examples, lemmas, open problems, propositions, templates, and theorems in order of occurrence, e.g., Convention 3.1, Template 3.2, and Example 3.3 in Chapter 3. The tables in Appendix A are named A.1 through A.5.

Table 1.6 suggests how to find items of particular interest.

1.5 Notes

In the Notes section ending each chapter, we mention supplementary reading on topics relevant to that chapter's theme. When works by several lead authors are cited, their names usually are ordered chronologically by year of first contribution.

The modern mathematical treatment of group choice, with its focus on formal evaluations of alternative consensus rules, began in the Enlightenment with contributions by Borda (1733–1799) [100], Condorcet (1743–1794) [137, 188, 256, 426], and their contemporaries. McLean's [253] survey of such work in 1784–1803 describes both axiomatic and probabilistic approaches to the design and analysis of voting procedures. McLean and London [257] identify aspects of Borda's and Condorcet's contributions that were anticipated in medieval works by Ramon Lull (c1235–1315) and Nicolas Cusanus (1401–1464). More generally, McLean and Urken [258] find contemporary issues of group choice in writings of Pliny the Younger (62?–c113), Lull, Cusanus, Borda, Condorcet, Lhuilier (1750–1840), Morales (c1790–1810), Daunou (1761–1840), Dodgson a.k.a. Lewis Carroll (1832–1898),

Table 1.6. *Finding Things*

To find	See
Additional reading	Notes sections on pages 8, 23, 51, 75, 100, 110
Axiomatic settings	Table 1.5
Axioms	Table A.3 on page 116; Index on page 147
Chapters	Contents on page vii
Cited references	Bibliography on page 119
Conventions	Table A.1 on page 114
Figures	List of Figures on page ix
Notations	Table A.2 on page 115
Open problems	Tables A.4 on page 117, A.5 on page 118
Sections	Contents on page vii
Tables	List of Tables on page xi

and Nanson (1850–1936). Black's [81, 85] history of the mathematical theory of committees and elections describes contributions of Borda, Condorcet, Laplace (1749–1827), Galton (1822–1911), Dodgson, and Nanson. McLean [255] analyzes Nanson's work in social choice and electoral reform. Suzumura [398] introduces the major lines of research in social choice theory and welfare economics during the twentieth century, research in part deeply influenced by Duncan Black [81, 85, 106, 346, 404] and Kenneth Arrow [11, 13, 18, 19, 20, 21].

The contemporary era of group choice began in 1948–1951 with Black's [73, 74, 75, 76, 77, 78, 79, 86] creation of a multidimensional spatial theory of voting and with Arrow's [10] formulation and analysis of the celebrated impossibility theorem. Black's 1958 monograph [81], which consolidated his early research and historical investigations, was reprinted in 1998 [85] along with later papers, e.g., [82, 83, 84]. Arrow's doctoral dissertation on the impossibility theorem appeared in 1951 as a monograph [11]; the second edition [13], now usually cited, appends a commentary entitled "Notes on the Theory of Social Choice, 1963." Of many subsequent monographs, those by Sen [372], Kelly [220], Fishburn [166], Campbell [119], and Aleskerov [4] have an appealing axiomatic stress. Accessible to nonspecialists are Barbut's [33] elementary introduction to social choice theory, Riker's [346] essay on the momentous contributions in the 1950s to social choice theory, Arrow's [14] views on formal theories of social choice, and Plott's [333] leisurely survey of axiomatic social choice theory. Sen [375], Pattanaik [322], and Campbell and Kelly [124] give more advanced reviews of social choice research in the Arrovian framework. Saari [356, 357, 358, 362], Tanguiane [401], and Stensholt [388] view Arrow's theorem and group choice from geometric perspectives. Moulin [302] and Barberà [32] stress the strategic theory of social choice, which concerns the conditions under which a sincere ballot is a voter's best strategy. Arrow, Sen, and Suzumura [20, 21] give a comprehensive introduction to social choice and welfare with, in particular, reviews on Arrovian impossibility theorems [5, 32, 124, 170], voting procedures [107], and the structure of social choice rules [323, 327].

In France the contemporary era of group choice began in 1952 with the publication of a paper by Guilbaud [190], which Arrow [13, p. 92] described as "a remarkable exposition of the theory of collective choice and the general problem of aggregation" and which helped to resurrect Condorcet's essay [137] "from the deep oblivion where it had fallen" [299]. Monjardet [293, 299] appraises the influence of Guilbaud's ideas on research in social choice theory, particularly [299] at Guilbaud's center in Paris, now called the Centre d'Analyse et de Mathématique Sociale at the École des Hautes Études en Sciences Sociales. Representative of this tradition until the early 1980s (and unavailable in English translation) are papers by Barbut [35, 36], Guilbaud and Rosenstiehl [192, 193], Feldman [160], Monjardet [289, 292], and Barthélemy [39, 40, 41].

Elementary logic, relations, and graphs are reviewed in most standard texts on discrete mathematics, e.g., Ross and Wright [352], while Davey and Priestley [148] provide an excellent introduction to ordered sets and lattices. More advanced treatments are by Suppes [395] for logic, Suppes [396] and Kaplansky [218] for set theory, Harary [202] and Berge [65] for graph theory, and Birkhoff [72], Crawley and Dilworth [142], and Grätzer [185] for lattice theory. Topics in bioinformatics are treated thoroughly by Stephen [389], Waterman [410], and Gusfield [195].

The theory of preference that Arrow uses ... is given the techniques bearing on the independence, consistency and completeness of an axiom system; and the change to an articulate mathematical symbolism well adapted to the material brought benefits of a kind and scale which, so far as the present author is concerned, could not have been foreseen. Its first fruits were a series of articles in the journals, some of them dealing with fundamental aspects of the theory of committees. By axiomatizing the theory Arrow's work had blown a sudden energy into the subject. — D. Black [84, p. 267] in 1972

Chapter 2

Axiomatics in Group Choice

When a group decision is at issue there is no existing means of analyzing the nature of the decision taken and of displaying the relation in which the decision stands to the opinions of the people by whom it is taken. — D. Black [73, p. 245] in 1948

Upon close examination, [my critics] implicitly accept the essential formulation stated here: The social choice from any given environment is an aggregation of individual preferences. The true grounds for disagreement are the conditions which it is reasonable to impose on the aggregation procedure, and even here it is possible to show that the limits of disagreement are not as wide as might be supposed from some of the more intemperate statements made. — K. J. Arrow [13, p. 103]

Several developments in group choice are the intellectual antecedents of recent axiomatic investigations of consensus problems in the biosciences. In section 2.1 we prove K. J. Arrow's impossibility theorem for weak orders, a result of outstanding significance in the theory of group choice. In section 2.2 we describe the structure of the decisive sets used to obtain that result. In section 2.3 we prove K. O. May's elegant possibility result, a characterization of the majority rule for weak orders on two alternatives. In later chapters we will apply the underlying paradigms to biological and data analysis problems having little to do with the theory of group choice.

Given sets of individuals (who vote) and alternatives (which individuals rank by preference), Arrow's basic premises are that

- The sets of individuals and of alternatives are fixed, finite, and unstructured.
- Each individual evaluates the alternatives by ranking them in a weak order.
- Each individual, when selecting a weak order, makes no attempt to manipulate the election's outcome.
- The individuals' weak orders can be aggregated into a weak order that is the consensus of the group.

We introduce weak orders using the notational conventions of Chapter 1.

Table 2.1. *Incidence Matrix of R_Z in Example* 2.1

	a	b	c	d	e	f
a	1	1	1	1	1	1
b	1	1	1	1	1	1
c	1	1	1	1	1	1
d	0	0	0	1	1	1
e	0	0	0	0	1	1
f	0	0	0	0	1	1

Table 2.2. *Ordered Partitions on $S_3 = abc$*

$a > b > c$	$a > bc$	$ab > c$	abc
$a > c > b$	$b > ac$	$ac > b$	
$b > a > c$	$c > ab$	$bc > a$	
$b > c > a$			
$c > a > b$			
$c > b > a$			

Example 2.1. Given a set $S = abcdef$ of alternatives, Bill has formed a *partition* $Y = \{abc, d, ef\}$ of S, i.e., a set of nonempty subsets of S, called *classes*, that are pairwise disjoint and that include every element of S. If two alternatives are in the same class, Bill is indifferent between them; if they are in distinct classes Bill strictly prefers one to the other and so he linearly orders the classes of Y to obtain an ordered partition $Z : abc > d > ef$. Ordered partitions are equivalent to weak orders as follows. From Z we can derive a binary relation R_Z on S by the following rule: for all $x, y \in S$, xy is in R_Z if x and y are in the same class of Z or if they are in distinct classes with the class of x preferred to the class of y. R_Z can be depicted by its incidence matrix (Table 2.1) or by the set $R_Z = \{aa, \ldots, cf, dd, de, df, ee, ef, fe, ff\}$ of its 25 ordered pairs. R_Z is complete (Table 1.2 on page 6) since if $x, y \in S$, then always xy $\in R_Z$ or yx $\in R_Z$. Transitivity (Table 1.2) holds since always $xy \in R_Z$ and $yz \in R_Z$ imply $xz \in R_Z$. Being complete and transitive, R_Z is a weak order (Table 1.3 on page 6). Since the whole argument reverses, a one-to-one correspondence exists between the sets of ordered partitions of S and weak orders on S: a problem on ordered partitions can be treated equivalently as a problem on weak orders. Table 2.2 lists the ordered partitions on S_3, from which the reader may list the weak orders on S_3. ■

Every weak order can be decomposed into useful subsidiary relations.

Definition 2.2. *For each weak order R on S let P be its* strict preference *relation*

$$(\forall x, y \in S)(xPy \iff \neg yRx),$$

so P is irreflexive and transitive; and let I be its indifference *relation*

$$(\forall x, y \in S)(xIy \iff xRy \wedge yRx),$$

so I is an equivalence relation whose classes are ordered by P.

Clearly P is a strict preference relation on S if and only if its complement $S^2 \setminus P$ is a weak order on S.

2.1 Impossibilities

While impossibility theorems, by themselves, do not provide a solution to the basic ethical problem of social choice, they do generate valuable insights and sharpen our ethical intuition in several ways. — P. K. Pattanaik [322, p. 201]

Our terminology for consensus problems on weak orders is reasonably standard. Let S be a set of n *alternatives*. Let k *individuals* participate in a process of collective decision making: each $i \in K = \{1, \ldots, k\}$ specifies an *individual order* R_i on S, where $R_i \in \mathcal{O} = \mathcal{O}_n$. The R_i form a *profile* $Q = (R_1, \ldots, R_k) \in \mathcal{O}^k$. From Q the decision-making process C derives a result $C(Q) \in \mathcal{O}$ called the *social order* for Q. C is a *social welfare function* by the following definition.

Definition 2.3. *Let a* partial function $f : \mathcal{O}^k \longrightarrow \mathcal{O}$ *be a binary relation* $R_f \subseteq \mathcal{O}^k \times \mathcal{O}$ *such that for each* $Q \in \mathcal{O}^k$ *at most one* $R \in \mathcal{O}$ *exists with* $(Q, R) \in R_f$. *A* social welfare function (SWF) *is a partial function* $C : \mathcal{O}^k \longrightarrow \mathcal{O}$, *where if* $Q = (R_1, \ldots, R_k) \in \mathcal{O}^k$, *then* $C(Q) = CQ = R$, *when it exists, is the social order for Q. To R_j and R there correspond strict preference relations, P_j and P, and indifference relations, I_j and I, where we*

shorten	*to*
$\lvert\{i \in K : xP_iy\}\rvert$	$N(xPy)$
$\lvert\{i \in K : xI_iy\}\rvert$	$N(xIy)$
$\lvert\{i \in K : xR_iy\}\rvert$	$N(xRy)$.

In a quite different context, where I is used to denote a subset of K, we

shorten	*to*
$(\forall i \in I \subseteq K)(xP_iy)$	xP_Iy
$\{i \in I \subseteq K : xR_iy\}$	$I_{xy}(Q)$.

If $I = K$, for example, then $K_{xy}(Q) = \{i \in K : xR_iy\}$.

SWFs, as well as consensus rules on equivalence relations or tree quasi-orders (Table 1.3 on page 6), are usually constrained so that the consensus problem is nontrivial.

Convention 2.4. *To any (multi)consensus rule C with domain $\mathcal{O}^k$ or $\mathcal{E}^k$ or $\mathcal{T}^k$ is associated a set $S = S_n$ of n alternatives on which the relations are defined. In this context, unless specifically stated otherwise, S is finite with $|S| = n \geq 3$.*

Table 2.3. *Axioms: Rules on Weak Orders. For notation see Definitions* 2.3 *and* 2.7.

***APO*: Anti-Pareto Optimality** $(\forall x, y \in S)(\forall Q \in \mathcal{O}^k)(x P_K y \Longrightarrow y P x)$

***Atn*: Autonomy** $(\forall x, y \in S)(\exists Q \in \mathcal{O}^k)(x R y)$

***CR*: Collective Rationality** $(\forall Q \in \mathcal{O}^k)(CQ$ is defined and single valued)

***Cst*$_1$: 1-Constant** $(\forall x, y \in S)(\forall Q \in \mathcal{O}^k)(x R y)$

***Dct*: Dictatorship** $(\exists j \in K)(\forall x, y \in S)(\forall Q \in \mathcal{O}^k)(x P_j y \Longrightarrow x P y)$

***DN*: Decisive Neutrality** $(\forall x, y, z, w \in S)(\forall Q, Q' \in \mathcal{O}^k)(K_{xy}(Q) = K_{zw}(Q') \wedge K_{yx}(Q) = K_{wz}(Q') \Longrightarrow [xRy \Longleftrightarrow zR'w] \wedge [yRx \Longleftrightarrow wR'z])$

***FT*: Free Triples**
$(\forall X = \{x, y, z\} \subseteq S)(\forall Q' \in \mathcal{O}^k)(\exists$ admissible $Q \in \mathcal{O}^k)(Q|_X = Q'|_X)$

***ID*: Inverse Dictatorship** $(\exists j \in K)(\forall x, y \in S)(\forall Q \in \mathcal{O}^k)(x P_j y \Longrightarrow y P x)$

***Ind*: Independence** $(\forall X \subseteq S)(\forall Q, Q' \in \mathcal{O}^k)(Q|_X = Q'|_X \Longrightarrow R|_X = R'|_X)$

***Ind*$_b$: Binary Independence**
$(\forall X = \{x, y\} \subseteq S)(\forall Q, Q' \in \mathcal{O}^k)(Q|_X = Q'|_X \Longrightarrow R|_X = R'|_X)$

***PO*: Pareto Optimality** $(\forall x, y \in S)(\forall Q \in \mathcal{O}^k)(x P_K y \Longrightarrow x P y)$

***PR*: Positive Responsiveness**
$(\forall x, y \in S)(\forall Q, Q' \in \mathcal{O}^k)((\exists j \in K)(([y P_j x \wedge x R'_j y] \vee [x I_j y \wedge x P'_j y]) \wedge (\forall i \in K \setminus \{j\})([x P_i y \Longrightarrow x P'_i y] \wedge [x I_i y \Longrightarrow x R'_i y])) \Longrightarrow [x R y \Longrightarrow x P' y])$

***Sym*: Symmetry** $(\forall Q \in \mathcal{O}^k)(\forall$ permutations σ of $K)$
$([CQ$ exists$] \wedge [C(R_{\sigma(1)}, \ldots, R_{\sigma(k)})$ exists$] \Longrightarrow CQ = C(R_{\sigma(1)}, \ldots, R_{\sigma(k)}))$

Consider what properties would be suitable to describe consensus rules on weak orders. We will formulate some axioms of SWFs (Table 2.3) and establish several relationships among them. To begin there is the dilemma that, depending on how a SWF C is defined, profiles Q may exist for which CQ is undefined. Consider, e.g., the majority rule.

Definition 2.5. *The method of* majority rule *for* $\mathcal{O}$ *is a SWF* $Maj : \mathcal{O}^k \longrightarrow \mathcal{O}$, *where*

$$(\forall x, y \in S)(\forall Q \in \mathcal{O}^k)(xy \in Maj Q \Longleftrightarrow N(xRy) \geq N(yRx)).$$

Is the value of *Maj* defined and single valued for every profile, i.e., is every profile *admissible* for *Maj*? An example with cyclic majorities shows the problem.

Example 2.6. *Paradox of Voting.* For $S = abc$ consider $Q = (R_1, R_2, R_3) \in \mathcal{O}^3$ with $R_1 = a > b > c$, $R_2 = b > c > a$, and $R_3 = c > a > b$. In these individual weak orders, count the occurrences of the ordered pairs $xy \in S^2$

	a	b	c
a	3	2	1
b	1	3	2
c	2	1	3

to see that $Maj\,Q$ should contain all and only the pairs in $\{aa, ab, bb, bc, ca, cc\}$. Since that relation is not transitive, it is not a weak order, so $Maj\,Q$ is undefined and Q is inadmissible for Maj. "Later, in working out an arithmetical example, an intransitivity arose, and it seemed to me that this must be due to a mistake in the arithmetic. On finding that the arithmetic was correct and the intransitivity persisted, my stomach revolted in something akin to physical sickness. Not only was the problem to which I had addressed myself more complicated than I had supposed, it was of a different kind." —Black [84, p. 262], recalling his discovery of the paradox in the 1940s. ■

The problem of inadmissible profiles can be addressed in various ways. One could force the SWF to be a function by imposing the axiom of collective rationality (*CR* in Table 2.3). Or one could exclude from consideration certain profiles of individual orders on a priori grounds; but the extent of such exclusions should be limited lest the problem become trivial. A way to ensure the robustness of a SWF involves restricting a relation on S to a subset of S as in the following definition.

Definition 2.7. *For each $X \subseteq S$ and $R \in \mathcal{R}$, let $R|_X = R \cap X^2$ be called the* restriction *of R to X. For each $Q = (R_1, \ldots, R_k) \in \mathcal{R}^k$, let $Q|_X = (R_1|_X, \ldots, R_k|_X)$ be the restriction to X of every $R_i \in Q$.*

If our a priori knowledge of the individual orders is incomplete to the extent that, for each set X of three alternatives, the weak order on X of every individual is completely unknown in advance, then it would seem to be inappropriate for some particular profile of weak orders on X *never* to occur by restriction from an admissible profile. Thus for every profile $Q' \in \mathcal{O}^k$ there should be at least one admissible profile $Q \in \mathcal{O}^k$ from which $Q'|_X$ can be obtained by restriction. Such is the motivation for the free-triples axiom (*FT* in Table 2.3). Clearly *CR* implies *FT*.

Pareto optimality (*PO* in Table 2.3) requires for all xy that if every individual strictly prefers x to y, then so must society. The axiom is named for Vilfredo Pareto [321], an Italian economist, mathematician, and sociologist whose work formed the foundation of modern welfare economics and whose ideas formed the basis of Italian fascism. But autonomy (*Atn* in Table 2.3) requires merely that the social order not be prevented a priori from having any ordered pair xy. Clearly *PO* implies *Atn*.

Let the members of a benevolent society conduct an election. Each member ranks a set S of alternatives. From the members' ballots the overall consensus ranking of S is calculated. When determining the consensus ranking of any subset X of S, $1 < |X| < |S|$, one should not have to take into account the members' rankings of any alternatives in $S \setminus X$.

Specifically, let two profiles of individual orders on S be such that when restricted to $X \subseteq S$ every individual's weak orders are identical. If the two social orders on S are then restricted to X, we would expect them also to be identical: relative to X the alternatives in $S \setminus X$ would be considered irrelevant. Independence of irrelevant alternatives (*Ind* in Table 2.3 on page 14) (Huntington [207], Arrow [10, p. 337], McLean [254]) is intuitively appealing, powerful, but not without disadvantages: "Indeed, when considering its various aspects one feels both an attraction and a repulsion to [Independence], wishing to adopt it but wishing also to reject it." — Black [84, p. 269].

Three axioms of Table 2.3 on page 14 are relevant to section 2.3. Decisive neutrality (*DN*) requires that if sets xy and zw of alternatives are used in the same way in profiles Q and Q', then the sets must be used in the same way in the social orders R and R'. Positive responsiveness (*PR*) requires that if the social order does not strictly prefer y to x and if the individual preferences remain the same except that one individual changes in a way favorable to x, then the new social order should strictly prefer x to y. Symmetry (*Sym*) requires that a SWF ensure the anonymity or equality of individuals: the social order should be determined only by the individual orders and not by the way the individuals (or subscripts) are associated with those orders.

Table 2.3 on page 14 has undesirable axioms. A 1-constant (Cst_1) SWF is uninformative since for every profile the social order is indifferent between every two alternatives. With anti-Pareto optimality (*APO*), if every individual prefers x to y, then the social order perversely prefers y to x. For a dictatorial (*Dct*) SWF, the social order is based on the preferences of one individual: if the dictator prefers x to y, then so must society. For an inverse dictatorship (*ID*), if the dictator prefers x to y, then the social order perversely prefers y to x. These axioms are related in basic ways.

Lemma 2.8. *Let $C : \mathcal{O}^k \longrightarrow \mathcal{O}$ be a SWF. Then*

$$CR \Longrightarrow FT, \tag{2.1a}$$

$$Ind \Longleftrightarrow Ind_b, \tag{2.1b}$$

$$DN \Longrightarrow Ind_b, \tag{2.1c}$$

$$DN \wedge PR \Longrightarrow PO \Longrightarrow Atn, \tag{2.1d}$$

$$Dct \vee ID \Longrightarrow \neg Sym. \tag{2.1e}$$

Proof. (2.1a), (2.1e), and the last part of (2.1d) easily follow from the definitions. *Concerning* (2.1b): If *Ind* holds, then independence holds for two-element subsets. If Ind_b holds, then independence holds for all two-element subsets of any given $X \subseteq S$ whence, by the restriction of weak orders, independence holds for X. *Concerning* (2.1c): Let $X = \{x, y\} \subseteq S$, $Q, Q' \in \mathcal{O}^k$ and $Q|_X = Q'|_X$, so that $K_{xy}(Q) = K_{xy}(Q')$ and $K_{yx}(Q) = K_{yx}(Q')$. Using *DN* with $z = x$ and $w = y$, then $xRy \Longleftrightarrow xR'y$ and $yRx \Longleftrightarrow yR'x$, whence $R|_X = R'|_X$, so that Ind_b holds. *Concerning* (2.1d): Let $Q \in \mathcal{O}^k$ be such that $(\forall i \in K)(xI_iy)$. Using *DN* with $Q' = Q$, $z = y$, and $w = x$, then xIy, so that $[N(xPy) = 0 \wedge N(yPx) = 0] \Longrightarrow xIy$. From this and *PR* we obtain $[N(xPy) = 1 \wedge N(yPx) = 0] \Longrightarrow xPy$; then induction and *PR* show that xPy for $N(xPy) = k$, whence *PO* holds. □

2.1.1 Arrow's Theorem

Finally, a series of abortive research ideas, each of which seemed to be more of a distraction than a help, culminated in my first major accomplishment, known as the theory of social choice. — K. J. Arrow [19, p. 46]

Social welfare could only be an aggregate of orderings. I already knew that majority voting, a plausible way of aggregating preferences, was unsatisfactory; a little experimentation suggested that no other method would work in the sense of defining an ordering. The development of the theorems and their proofs then required only about three weeks, although writing them as a monograph (Social Choice and Individual Values) *took many months.* — K. J. Arrow [15, p. 4]

The import of Arrow's impossibility theorem for weak orders is that dictatorships, which are undesirable, are a consequence of desirable properties.

Theorem 2.9 (Arrow). [13, p. 97]. *Let $C : \mathcal{O}^k \longrightarrow \mathcal{O}$ be a SWF. C satisfies Dct if it satisfies FT, Ind, and PO; C satisfies ID if it satisfies FT, Ind, and APO.*

Our proof of Arrow's theorem uses sets of individuals who are decisive in the sense that by acting together they could influence an election's result.

Definition 2.10. *Given a SWF C, let $I \subseteq K$ and $x, y \in S$ with $x \neq y$. I is called* decisive for xy, *a condition denoted by U^I_{xy}, if every $Q = (R_1, \ldots, R_k) \in \mathcal{O}^k$ having xP_Iy then implies xPy. I is called* decisive *if it is decisive for all $x, y \in S$ with $x \neq y$. U_C is the set of all decisive sets. I is called* inversely decisive for xy, *a condition denoted by V^I_{xy}, if every $Q \in \mathcal{O}^k$ having xP_Iy then implies yPx. I is called* inversely decisive *if it is inversely decisive for all $x, y \in S$ with $x \neq y$. V_C is the set of all inversely decisive sets.*

Example 2.11. Let $C : \mathcal{O}^k \longrightarrow \mathcal{O}$ be a dictatorial SWF with $K = 123$ and 1 as the dictator; then $U_C = \{1, 12, 13, 123\}$ and $V_C = \emptyset$. ■

Decisive sets for weak orders exhibit invariance properties, which are based on a technical property of binary relations.

Lemma 2.12. [92, p. 65]. *Let D be an irreflexive relation on S such that $aDb \Longrightarrow aDx \wedge xDb$ for all $x \in S$ except where irreflexivity forbids. Then $aDb \Longrightarrow xDy$ for all $x, y \in S$ with $x \neq y$.*

Proof. Imagine D on a grid S^2 of points in the plane. The hypothesis asserts that if D has a point ab, it also has points of the horizontal and vertical lines through ab except for points on the diagonal $x = y$. Consider $x, y \in S$ with $x \neq y$. If $a \neq y$, then $aDb \Longrightarrow aDy \Longrightarrow xDy$. If $b \neq x$, then $aDb \Longrightarrow xDb \Longrightarrow xDy$. If $x = b$ and $y = a$, then $c \in S$ exists with $c \neq a$ and $c \neq b$, since $|S| \geq 3$, whence $aDb \Longrightarrow aDc \Longrightarrow bDc \Longrightarrow bDa = xDy$. □

Lemma 2.13 establishes an invariance requirement for $I \subseteq K$: for all $xy, ab \in S^2$ with $x \neq y$ and $a \neq b$, it demands that I be decisive for xy if and only if I is decisive for ab. This requirement, which Sen [374] calls *invariant decisiveness*, prevents the use of any information regarding particular features of alternatives when discriminating among them.

Lemma 2.13. [374, p. 33]. *Let $C : \mathcal{O}^k \longrightarrow \mathcal{O}$ be a SWF that satisfies FT, Ind, and Atn. If $I \subseteq K$, then*

Direct: $(\exists ab \in S^2)(U^I_{ab}) \Longleftrightarrow I \in U_C$.

Inverse: $(\exists ab \in S^2)(V^I_{ab}) \Longleftrightarrow I \in V_C$.

Proof. Assume U^I_{ab} for $ab \in S^2$. We first prove that $U^I_{ab} \Longrightarrow U^I_{ax}$ for all $x \neq a$. If $x = b$, U^I_{ax} is true by hypothesis. If $x \neq b$, let $Q \in \mathcal{O}^k$ have aP_Ib and aPb, and since *Atn* holds, let $Q' \in \mathcal{O}^k$ have $bR'x$. Using *Ind* and *FT*, let $Q'' \in \mathcal{O}^k$ have aP''_Ib, aP''_Ix, and $Q''|_{\{b,x\}} = Q'|_{\{b,x\}}$; i.e., in Q' move a without changing the $\{b, x\}$-configurations. Then $aP''b \wedge bR''x \Longrightarrow aP''x$, whence U^I_{ax}. Similarly, $U^I_{ab} \Longrightarrow U^I_{xb}$ for all $x \neq b$. By Lemma 2.12 on the page before, $U^I_{ab} \Longrightarrow U^I_{xy}$ for all $xy \in S^2$ with $x \neq y$, whence $I \in U_C$. The converse is trivial. The inverse result follows similarly. $\square$

Lemma 2.14 establishes an invariance requirement for $I, J \subseteq K$: it demands that I and J have the same status, with respect to decisiveness, if J contains I and $J \setminus I$ is not decisive. This requirement, which Sen [374] calls *equivalent subsets*, prevents the use of any information regarding the presence or absence of individuals who themselves do not form a decisive subset.

Lemma 2.14. [374, pp. 33–34]. *Let $C : \mathcal{O}^k \longrightarrow \mathcal{O}$ be a SWF that satisfies FT, Ind, and Atn. If I and J satisfy $I \subseteq J \subseteq K$, then*

Direct: $J \setminus I \notin U_C \Longrightarrow [I \in U_C \Longleftrightarrow J \in U_C]$.

Inverse: $J \setminus I \notin V_C \Longrightarrow [I \in V_C \Longleftrightarrow J \in V_C]$.

Proof. Assume $J \setminus I \notin U_C$. Clearly $I \in U_C \Longrightarrow J \in U_C$, so instead let $J \in U_C$. Using *Ind* and *FT* let $Q \in \mathcal{O}^k$ have xP_Iy, xP_Iz, $xP_{J\setminus I}y$, and $zP_{J\setminus I}y$; then xPy since $J \in U_C$. If zPy, then $U^{J\setminus I}_{zy}$, so that $J \setminus I \in U_C$ by Lemma 2.13, which is false; thus $yRz \wedge xPy \Longrightarrow xPz$, so that U^I_{xz}, whence $I \in U_C$ by Lemma 2.13. The inverse result follows similarly. $\square$

If any nonsingleton set $J \subseteq K$ is decisive, then Lemma 2.14 allows us to partition J into strictly smaller parts, I and $J \setminus I$, where I is again decisive. Choose any set $L \subset J$ such that $1 \leq |L| < |J|$: if L is decisive, then put $I = L$, whence I is decisive and has $1 \leq |I| < |J|$; if L is not decisive, then put $I = J \setminus L$, whence I is decisive by Lemma 2.14 and has $1 \leq |I| < |J|$. By recursively applying this partitioning process to the smaller decisive part, we eventually obtain a decisive singleton set $\{i\}$ for some $i \in K$. Let this procedure be called *recursive partitioning* by (in this case) equivalent subsets; with it we obtain this proof.

Table 2.4. *Sen's Strategy to Prove Impossibility*

[374, pp. 32–34]. Let C be a consensus rule on $\mathcal{X}$ for which axioms of independence, Pareto optimality, and dictatorship are specified.

1. For sets $I \subseteq K$, define a concept of decisiveness.
2. For pairs $\alpha, \beta \in S^m$ of m-tuples, establish a property of invariant decisiveness.
3. For pairs $I, J \subseteq K$, establish a property of equivalent subsets (or equivalent).
4. Use these properties to prove by recursive partitioning that independence and Pareto optimality imply dictatorship.

***Proof of Arrow's Theorem* 2.9 on page 17.** Let $C : \mathcal{O}^k \longrightarrow \mathcal{O}$ be any SWF that satisfies *FT*, *Ind*, and *PO*. Since K is decisive (by *PO*) and finite, recursive partitioning by equivalent subsets (using Lemma 2.14) shows that $\{i\} \in U_C$ for some $i \in K$, whence *Dct* holds. Similarly, *ID* holds if C satisfies *FT*, *Ind*, and *APO*. □

Table 2.4 is a general strategy (due to Sen) for proving impossibility results. To prove Arrow's theorem with it, take $m = 2$ and let Definition 2.10 on page 17 and Lemmas 2.13 and 2.14 provide the concepts and properties in steps 1–3.

2.1.2 Wilson's Theorem

[T]he essential significance of Arrow's Theorem is not diminished if one abandons the Pareto Principle. The theorems that we obtain here are, of course, somewhat weaker than Arrow's Theorem, but the fact remains that Arrow's other conditions suffice to exclude all of the democratic social choice processes of interest.
— R. B. Wilson [420, p. 478]

Arrow's theorem can be stated more generally, for if Pareto optimality is weakened to autonomy, then C still has undesirable properties.

Theorem 2.15 (Wilson). [420, p. 484]. *Let $C : \mathcal{O}^k \longrightarrow \mathcal{O}$ be a SWF. If C satisfies Atn, FT, and Ind, then it satisfies either Cst_1, Dct, or ID.*

Unless such a SWF is 1-constant, it has at least one (inversely) decisive set.

Lemma 2.16. *Let $C : \mathcal{O}^k \longrightarrow \mathcal{O}$ be a SWF. If C satisfies Atn, $\neg Cst_1$, FT, and Ind, then either $K \in U_C$ or $K \in V_C$.*

Proof. If C is not 1-constant, then xPy for some $x, y \in S$ and $Q \in \mathcal{O}^k$. Choose $z \in S \backslash \{x, y\}$ and $Q' \in \mathcal{O}^k$ arbitrarily except that $xP'_K z$, $yP'_K z$, and $Q'|_{\{x,y\}} = Q|_{\{x,y\}}$. Then $xP'y$ by *Ind*. If $yR'z$, then $xP'z$ by transitivity, so U^K_{xz} by *Ind*, whence $K \in U_C$ by Lemma 2.13. If $zP'y$, then V^K_{yz} by *Ind*, whence $K \in V_C$ by Lemma 2.13. □

With Lemma 2.14 on page 18 and Lemma 2.16 on page 19 we obtain this proof.

***Proof of Wilson's Theorem* 2.15.** Let $C : \mathcal{O}^k \longrightarrow \mathcal{O}$ be any SWF that satisfies *Atn*, *FT*, and *Ind*. If $K \in U_C$, then since K is finite, recursive partitioning by equivalent subsets (using Lemma 2.14 on page 18) shows that $\{i\} \in U_C$ for some $i \in K$, whence *Dct* holds. Similarly, *ID* holds if $K \in V_C$. If $K \notin U_C \cup V_C$, then Cst_1 holds by Lemma 2.16. □

Arrow's theorem follows easily from Wilson's theorem.

***Proof of Arrow's Theorem* 2.9 on page 17.** Let $C : \mathcal{O}^k \longrightarrow \mathcal{O}$ be any SWF that satisfies *FT*, *Ind*, and *PO*. Since $PO \Longrightarrow Atn$, then one of $\{Dct, ID, Cst_1\}$ holds by Theorem 2.15, but since *PO* is inconsistent with *ID* and Cst_1, only *Dct* remains. □

By Wilson's theorem a significant class of SWFs can be partitioned into those that are dictatorial, those that are inversely dictatorial, and those that are 1-constant. That class can be partitioned also into those that are Pareto optimal, those that are anti-Pareto optimal, and those that are 1-constant.

Theorem 2.17 (Malawski and Zhou). [243, p. 104]. *Let $C : \mathcal{O}^k \longrightarrow \mathcal{O}$ be a SWF. If C satisfies Atn, FT, and Ind, then it satisfies APO, Cst_1, or PO.*

Proof. Let $C : \mathcal{O}^k \longrightarrow \mathcal{O}$ be any SWF that satisfies *Atn*, *FT*, and *Ind*. If Cst_1 holds, we are almost done. Otherwise $K \in U_C \cup V_C$ by Lemma 2.16, but then *PO* holds when $K \in U_C$ and *APO* holds when $K \in V_C$. Since *APO*, Cst_1, and *PO* are pairwise inconsistent, C satisfies exactly one of these properties. □

With suitable domain restrictions, independence and autonomy imply Pareto optimality. Malawski and Zhou's theorem is not an impossibility theorem since a desirable property (*PO*) is a consequence of its hypotheses. In its presence, Wilson's theorem follows easily from Arrow's theorem, whence the two are equivalent in the sense that each implies the other.

***Proof of Wilson's Theorem* 2.15.** Let SWF $C : \mathcal{O}^k \longrightarrow \mathcal{O}$ satisfy *Atn*, *FT*, and *Ind*. By Theorem 2.17, exactly one of $\{APO, Cst_1, PO\}$ holds. By Theorem 2.9 on page 17, $PO \Longrightarrow Dct$ and $APO \Longrightarrow ID$, so exactly one of $\{Cst_1, Dct, ID\}$ holds. □

2.1.3 Sen's Theorem

The set $\{CR, DN, PR, Sym\}$ lies in the twilight zone between possibility and impossibility: for weak orders on two alternatives it is consistent and characterizes the method of majority rule (section 2.3), but for weak orders on more than two alternatives it is inconsistent.

Theorem 2.18 (Sen). [372, p. 73]. *Let $C : \mathcal{O}^k \longrightarrow \mathcal{O}$ be a SWF. For C the set $\{CR, DN, PR, Sym\}$ is inconsistent.*

Proof. Using (2.1a)–(2.1d) on page 16, $CR \wedge DN \wedge PR \Longrightarrow FT \wedge Ind \wedge PO$, whence *Dct* holds by Arrow's Theorem 2.9 on page 17, so *Sym* does not hold by (2.1e) on page 16. □

2.2 Decisiveness

Followers of Bourbaki will notice an ultrafilter in the background.
— J. H. Blau [93, p. 202]

Invariance relationships for decisive sets are basic to proving impossibility results for SWFs. While the relationships of invariant decisiveness (Lemma 2.13 on page 18) and equivalent subsets (Lemma 2.14 on page 18) reveal structure in U_C and V_C, there is more. Clearly every superset of a decisive set is decisive by Definition 2.10 on page 17; although every subset of a decisive set need not be decisive, the set intersection of decisive sets is decisive, whence U_C exhibits a relationship of *intersection invariance*.

Lemma 2.19. [119, p. 47]. *Let* $C : \mathcal{O}^k \longrightarrow \mathcal{O}$ *be a SWF that satisfies FT, Ind, and Atn. If* $I, J \subseteq K$, *then*

Direct: $I \cap J \in U_C \Longleftrightarrow I, J \in U_C$.

Inverse: $I \cap J \in V_C \Longleftrightarrow I, J \in V_C$.

Proof. Let $I \cap J \in U_C$, whence $\bigcap_{i \in I \cap J} P_i \subseteq P$ for all $Q \in \mathcal{O}^k$. Since $I \cap J \subseteq I$, then $\bigcap_{i \in I} P_i \subseteq \bigcap_{i \in I \cap J} P_i$ for all $Q \in \mathcal{O}^k$. Thus $\bigcap_{i \in I} P_i \subseteq P$ for all $Q \in \mathcal{O}^k$, whence $I \in U_C$, and similarly for J. Next let $I, J \in U_C$. Let x, y, z be any three distinct members of S and choose $Q \in \mathcal{O}^k$ arbitrarily except that $x P_I y$ and $y P_J z$. Since $I, J \in U_C$, then $x P y$ and $y P z$, whence $x P z$ by transitivity of R. Then $U_{xz}^{I \cap J}$ since the choice of Q is consistent with any ordering of x and z for individuals $i \in K \setminus (I \cap J)$, whence $I \cap J \in U_C$ by Lemma 2.13 on page 18. The inverse result follows similarly. □

If $J \in U_C$ and $I \subseteq J$, then I and $J \setminus I$ have unlike statuses with respect to decisiveness and exhibit an invariance relationship of *unlike complements*.

Lemma 2.20. [119, p. 55]. *Let* $C : \mathcal{O}^k \longrightarrow \mathcal{O}$ *be a SWF that satisfies FT, Ind, and Atn. If* I *and* J *satisfy* $I \subseteq J \subseteq K$, *then*

Direct: $J \in U_C \Longrightarrow [I \in U_C \Longleftrightarrow J \setminus I \notin U_C]$.

Inverse: $J \in V_C \Longrightarrow [I \in V_C \Longleftrightarrow J \setminus I \notin V_C]$.

Proof. Choose any pairwise distinct $x, y, z \in S$, with $Q \in \mathcal{O}^k$ such that $x P_J y$, $x P_I z$, and $z P_{J \setminus I} y$. Then $x P y$ because $J \in U_C$. If $x P z$, then U_{xz}^{I} by *Ind* and the fact that Q is consistent with any ordering of x and z by individuals $i \in J \setminus I$, whence $I \in U_C$ by Lemma 2.13 on page 18. If $z R x$, then $z P y$ by transitivity, so $U_{zy}^{J \setminus I}$ by *Ind* and the fact that

Q is consistent with any ordering of y and z by individuals $i \in I$, whence $J \setminus I \in U_C$ by Lemma 2.13, as required. The inverse result follows similarly. □

With Lemma 2.20 we obtain this proof.

***Proof of Arrow's Theorem* 2.9 on page 17.** Let $C : \mathcal{O}^k \longrightarrow \mathcal{O}$ be any SWF that satisfies *FT*, *Ind*, and *PO*. Since K is decisive (by *PO*) and finite, recursive partitioning by unlike complements (using Lemma 2.20) shows that $\{i\} \in U_C$ for some $i \in K$, whence *Dct* holds. Similarly, *ID* holds if C satisfies *FT*, *Ind*, and *APO*. □

Lemma 2.16 on page 19 and Lemmas 2.19 and 2.20 on page 21 also yield a proof.

***Proof of Wilson's Theorem* 2.15 on page 19.** [119, p. 56]. Let $C : \mathcal{O}^k \longrightarrow \mathcal{O}$ be any SWF that satisfies *FT*, *Ind*, and *Atn*. For $t \in K$ set $K_t = \{i \in K : i \neq t\}$. Let $K \in U_C$. If $\{i\} \in U_C$ for some $i \in K \setminus \{k\}$, then *Dct* holds; else $K_i \in U_C$ for all $i \in K \setminus \{k\}$ by Lemma 2.20, so that $\{k\} = \bigcap_{i \in K \setminus \{k\}} K_i \in U_C$ by Lemma 2.19 and finiteness of K, whence *Dct* holds. Similarly, *ID* holds if $K \in V_C$. If $K \notin U_C \cup V_C$, then Cst_1 holds by Lemma 2.16 on page 19. □

The families U_C and V_C are set-theoretic structures that are also used in topology [103] and model theory [63].

Definition 2.21. Bourbaki [103, vol. 1, pp. 57–68]. *A set $F \subseteq 2^K$ is called a* filter on K *if*

$$\emptyset \notin F \wedge K \in F, \tag{2.2a}$$

$$(\forall I \in F)(\forall J \subseteq K)(I \subseteq J \Longrightarrow J \in F), \tag{2.2b}$$

$$(\forall I, J \in F)(I \cap J \in F). \tag{2.2c}$$

F is called an ultrafilter *if also*

$$(\forall I \subseteq K)(K \setminus I \notin F \Longrightarrow I \in F). \tag{2.2d}$$

Lemmas 2.16, 2.19, and 2.20 show that families of decisive sets are ultrafilters.

Theorem 2.22. *Let $C : \mathcal{O}^k \longrightarrow \mathcal{O}$ be a SWF. If C satisfies FT, Ind, Atn, and $\neg Cst_1$, then either U_C or V_C is an ultrafilter.*

Proof. By Lemma 2.16 on page 19, assume $K \in U_C$ and prove that U_C is an ultrafilter. Lemma 2.19 establishes (2.2b) and (2.2c) for U_C. In Lemma 2.20 take $J = K$ to see that (2.2d) holds for U_C, and $I = J = K$ to see that (2.2a) holds for U_C. □

The converse also holds.

Theorem 2.23. [200, p. 93]. *If F is an ultrafilter on K, then there exists a SWF $C : \mathcal{O}^k \longrightarrow \mathcal{O}$ satisfying CR, Ind, and PO such that $U_C = F$.*

To see this, define C by $(\forall x, y \in S)(\forall Q \in \mathcal{O}^k)(xCQy \Longleftrightarrow K_{xy}(Q) \in F)$.

2.3 Possibilities

In this section let $|S| = 2$. In contrast to Sen's Theorem 2.18 on page 20, the axioms *CR*, *DN*, *PR*, and *Sym* now characterize *Maj*, the method of majority rule (Definition 2.5 on page 14).

Theorem 2.24 (May). [249, p. 682]. *Let $C : \mathcal{O}^k \longrightarrow \mathcal{O}$ be a SWF on $S = xy$. $C = Maj$ if and only if C satisfies CR, DN, PR, and Sym.*

Proof. Let $C = Maj$. *CR* holds since *Maj* always gives a unique result. *DN* holds since the definition of *Maj* is unaffected by interchanging x and y. *PR* holds since changing one individual's preference breaks a tie. *Sym* holds since a *Maj* result for $S = xy$, being determined by $N(xPy)$ and $N(yPx)$, is independent of which individuals hold these preferences. Conversely let $Q \in \mathcal{O}^k$ and let C satisfy *CR*, *DN*, *PR*, and *Sym*. By *Sym*, CQ must depend only on $N(xPy)$, $N(xIy)$, and $N(yPx)$. By *DN*, $(N(xPy) = N(yPx) \Longrightarrow xIy)$ as can be checked by assuming the contrary and permuting x and y in each individual's weak order. Using this and *PR* we obtain $(N(xPy) > N(yPx) \Longrightarrow xPy)$ and $(N(xPy) < N(yPx) \Longrightarrow yPx)$, whence $C = Maj$. □

Thus with two alternatives, any consensus rule other than the majority rule will favor one alternative over the other, favor one individual over another, fail to give a definite result for some profile, or fail to respond positively to changes in individual preferences. Also the four axioms are logically independent [250] in the sense that nothing is inconsistent about any combination of their truth values.

2.4 Notes

Brams and Fishburn [107] survey voting procedures that aggregate individuals' preferences to obtain collective decisions. Campbell and Kelly [124] investigate Arrow's impossibility theorem and search for reasonable social choice rules by relaxing constraints within the Arrovian axiomatic framework. Aleskerov [5] develops different types of voting schemes that make sense within the Arrovian framework. Since collective rationality (unrestricted domain) is often assumed by impossibility theorems, Gaertner [170] explores the various ways in which domain restrictions can be relaxed. Although we emphasize Arrow's impossibility theorem, social choice theory has other well-known impossibility results, e.g., the impossibility of a Paretian liberal (Sen [372, p. 87], [373]) and the impossibility of strategy-proof voting procedures (Gibbard [178], Satterthwaite [366], Gärdenfors [173], Barberà [31, 32]).

For section 2.1 (Impossibilities): The paradox of voting (see Example 2.6), also called the Condorcet effect (Guilbaud [190]) or the paradox of cyclical majorities (Black

[81]), has been studied extensively, e.g., Black [75], Greenberg [187], Plott [332], Weber [412], Jones et al. [214], and especially Gehrlein [176]. The basic impossibility results we report are by Arrow [13], Wilson [420], and Malawski and Zhou [243]. Monjardet [290] establishes Arrovian impossibility results for tournaments (complete, antisymmetric binary relations). Suzumura [397] gives a systematic presentation of Arrovian impossibility theorems and analyzes the simple majority rule as a collective choice mechanism. Fishburn [166] reviews contributions to social choice theory that are based on Arrow's approach and subsequent developments. Barthélemy [45] reviews aggregation procedures in group choice, emphasizing combinatorial and algorithmic aspects. Campbell [119] establishes the classical impossibility theorems under restrictions typically assumed for resource allocation models. Kelly [220] and Aleskerov [4] emphasize Arrovian impossibility theorems, while Moulin [301] explores their relevance to game theory.

Although his original statement [10, 11, 12] of the impossibility theorem uses axioms of positive association and citizens' sovereignty, Arrow [13, p. 97] later replaces them by Pareto optimality. Blau [89] and Murakami [307] correct an error in the original statement [10, 11, 12]. Arrow's 1963 version [13, p. 97] of the theorem incorporates these changes and is the one usually cited. Although our proof of Theorem 2.9 on page 17 is based on decisive voters (Arrow [13], Blau [92], Sen [374]), proofs can be based on pivotal voters (Barberà [30]), extremely pivotal voters (Geanakoplos [175]), or topological (Baryshnikov [56, 57]), geometric (Saari [359]), or Fourier-theoretic (Kalai [215]) concepts. Researchers have explored the general nature of Arrow's theorem by modifying its codomain, e.g., Sen [371], Schick [367], Hansson [198, 200], Nakamura [309, 310, 311], Blair and Pollak [88], Blau [93], and Monjardet [295]; by permitting an infinite number of voters, e.g., Fishburn [163], Kirman and Sondermann [226], Hansson [200], Schmitz [368], Armstrong [8], Monjardet [295], Fishburn [166, chap. 10], Campbell [119, chap. 9], Chichilnisky and Heal [133], and McMorris and Powers [273]; or by taking continuous or topological approaches to the study of social choice, e.g., Kelly [219], Saposnik [365], Chichilnisky [128, 129, 130, 131], Chichilnisky and Heal [132], Campbell [119], Heal [203], Lauwers [232], and Baigent [24]. Arkhipoff [7] uses category theory to develop an axiomatic theory of aggregation based on Arrow's theorem. Hansson [201] extends the formal framework of social choice theory by introducing separate representations of preferences and choices. Campbell and Kelly [125] investigate the relationships on different domains between Arrow's and Wilson's impossibility theorems, while generalizations of Wilson's [420] theorem for weak orders are obtained by Monjardet [290] for tournaments and by McMorris and Powers [269] for unrooted phylogenies. Huntington [207], Hansson [197, 199], Blau [91], Ray [343], Campbell [115], Bordes and Tideman [102], McLean [254], and Cramer-Benjamin [140] examine implications or interpretations of the powerful independence axiom. Campbell and Kelly [120, 121] and Powers [336, 337] investigate the structure of the set of alternatives for independent consensus rules. Monjardet [296], Crown, Janowitz, and Powers [144, 145, 146], Leclerc and Monjardet [240], and Sholomov [379] study the implications of neutrality in general mathematical models of consensus. The extent to which one can modify axioms in impossibility formulations, while retaining impossibility, has been investigated by many authors, e.g., Mas-Colell and Sonnenschein [247], Fishburn [165], Blair et al. [87], Baigent [23], Kelly [221], Campbell and Kelly [123, 124], and Powers [338]. If consensus rules on weak orders are allowed to return more than one result, Vincke [406] and Bouyssou [105] show that a consensus rule can be independent and Pareto optimal, yet not be dictatorial in the usual strong sense.

For section 2.2 (Decisiveness): Sen [374] and Campbell [119] provide the basis for our treatment of families of decisive sets for weak orders. Brown [110] investigates Arrovian SWFs via the lattice obtained by ordering families of decisive sets by set inclusion. Authors using filters or ultrafilters to analyze the structure of social choice functions include Kirman and Sondermann [226], Brown [110, 111], Hansson [200], and Monjardet [290, 291, 295]. For relationships between simple games (Shapley [377]) and social choice theory see Guilbaud [190], Blau [90], Wilson [419], Bloomfield and Wilson [95], Nakamura [309, 310, 311], Bloomfield [94], and Peleg [326]. For connections among simple games, ultrafilters, and aggregation rules that are stable (nonmanipulable, strategy-proof) in the sense used by the fundamental impossibility result of Gibbard [178] and Satterthwaite [366], see Pazner and Wesley [324], Ishikawa and Nakamura [208], and Batteau, Blin, and Monjardet [60].

For section 2.3 (Possibilities): May [249, 250] obtained the first axiomatic characterization of the majority rule SWF, but see related work on the majority rule by Sen [370], Bordes [101], Straffin [393], Campbell [116, 117, 118], Nitzan and Paroush [317], Maskin [248], Campbell and Kelly [122], and Regenwetter, Marley, and Grofman [344]. Other consensus rules on weak or linear orders have been characterized: see Smith [382], Fishburn [164], Young [423, 424], and Myerson [308] for scoring rules in general; Gärdenfors [172], Young [422], Fishburn and Gehrlein [167], Nitzan and Rubinstein [318], Saari [355], and Marchant [244] for the Borda rule [100]; Richelson [345], Roberts [347, 348], and Saari [359] for the plurality rule; Young and Levenglick [427] and Young [426] for the median rule (also called the Condorcet [297] or Kemeny [222, 224] rule), which is shown by Monjardet [297] to have been rediscovered many times. Michaud [278], Young [426], and Monjardet [297] give derivations of the Condorcet rule from Condorcet's writings; on this topic the explication of his writings is challenging (McLean and Hewitt [256], McLean [253]). Gehrlein [176] reviews the research on estimating for various voting procedures the probability of a Condorcet winner (a candidate that would defeat each other candidate by majority rule in a series of pairwise elections) or the Condorcet efficiency (the conditional probability that a procedure elects the Condorcet winner, given that a Condorcet winner exists); Gehrlein and Lepelley [177] provide a representative example of this literature. Pattanaik [323] reviews the positional voting rules [167, 172, 354, 361], which exploit the position of each alternative within each individual order and which include various Borda and scoring rules; for nonpositional (pairwise) rules, see Saari [360].

Apart from the difficulty of understanding Mathematics, which everyone feels and some people feel acutely, there is the drawback that much of the existing Mathematics was developed to deal with physical problems and is not well adapted to deal with the human sciences. In time a new Mathematics will be invented.
— D. Black [80, p. 513] in 1950

A recurring theme is the arbitrariness of what we choose to regard as a proper explanation and the associated clash of cultures between mathematics and biology. In general, mathematicians value conceptual simplicity and the idealized model of a process, whereas biologists want to know how the specific system they are confronting actually works.
— J. Slack [381] in 2002

Chapter 3
Impossibilities in Bioconsensus

It is a widespread fallacy that what mathematics contributes to biology is quantification of an otherwise innumerate science. But experimental biologists have long been expert at measuring and quantifying. The real contribution of mathematics lies in a precise qualitative framework of reasoning. — C. R. M. Bangham and B. Asquith [29]

As such [the axiomatic method] is not a new invention; but its systematic use as an instrument of discovery is one of the original features of contemporary mathematics.
— N. Bourbaki [104, p. 8]

In 1951 Arrow's book stimulated controversy and research in social choice theory concerning logical restrictions on ways to aggregate individual preferences into social preferences (see Kelly [220]). But although aggregation models arise in many areas of science and technology, 24 years elapsed before Arrovian results began to appear outside of social choice theory. In 1975 Wilson [421, p. 89] encouraged the extension of Arrow's axiomatic approach: "It is natural to ask whether procedures for aggregating attributes other than preferences are subject to similar restrictions." And in 1975 Mirkin [282] published an impossibility theorem for aggregating partitions of a set, a problem relevant to cluster and data analysis since partitions of a set model nonhierarchical classifications. After discussing Mirkin's result (section 3.1), we will describe (sections 3.2–3.5) impossibility theorems for aggregating hierarchical structures of interest to bioconsensus in general and classification in particular. We consider several ways to represent treelike information structures on a set S. The tree structure can be viewed as a collection of n-ary associations among the elements of S, e.g., as a binary relation of ancestry (section 3.2) or as higher order relations of proximity (sections 3.3 and 3.4), or it can be viewed as a hypergraph [241], which is simply a set of subsets of S (sections 3.4 and 3.5). For all these cases we will investigate whether the intended properties of such structures lead to Arrovian impossibility theorems. When aggregating such structures, one usually takes consensus rules to be collectively rational.

Convention 3.1. *In Chapter 3 and thereafter, unless specifically stated otherwise, each (multi)consensus rule C is assumed to be collectively rational, i.e., $C(P)$ is defined and single valued for every profile P.*

Thus the impossibility theorems become minor variants of the following template.

Template 3.2. *If a consensus rule satisfies Ind and PO, then it satisfies Dct.*

In social choice theory dictatorial consensus rules may be unsuitable since they violate democratic principles, but in data analysis their use may be appropriate.

Example 3.3. [283, p. 127]. If individuals represent factors influencing the result of an analysis of data, then a dictatorial consensus rule simply chooses a particular factor on which to base all subsequent conclusions. ■

Readers may already have encountered the protagonist in the following example.

Example 3.4. [52, p. 59]. Through years of experience a user becomes convinced that a particular hierarchical clustering algorithm (A_1) will almost certainly produce meaningful clusters after operating on any real data set worth analyzing. But the user is willing to refine the output of A_1 with the outputs of other hierarchical algorithms ($A_2, \ldots, A_k$). Let these algorithms produce a profile $P = (H_1, \ldots, H_k)$ of hierarchies. The user requires that every cluster of H_1 be in the final hierarchy CP, but scans for clusters of the other H_i that are both mutually consistent and consistent with all the clusters of H_1. CP is then the hierarchy that contains all these mutually consistent clusters. Although this procedure may be plausible, it is dictatorial since $H_1 \subseteq CP$. ■

A dictator may be *strong*, by preventing all other individuals from contributing to the consensus result, or *weak*, by allowing individuals to affect the consensus result as long as they do not contravene the dictator's preferences. In political terms a strong dictatorship is "the most corrupt form of government because it serves only the base desires of the ruler and ignores the counsel of the wise" [242, p. 28]. Where dictatorships are inevitable, their characterization as strong or weak is desirable.

3.1 Partitions of a Set

Classification investigates sets of elements to decide if they can be summarized validly in terms of a small number of classes of similar elements. Consider decomposing a set S into a nonhierarchical classification or partition Y of S (defined in Example 2.1 on page 12). Recall the well-known one-to-one correspondence between the set of partitions of S and the set of equivalence relations on S.

Definition 3.5. *For each equivalence relation R on S, if $x \in S$, then $\bar{x} = \{y \in S : xRy\}$ is called an* equivalence class *of R and $Y_R = \{\bar{x} : x \in S\}$ is the partition of S corresponding to R. Conversely, for each partition Y of S then $R_Y = \{xy \in S^2 : (\exists Z \in Y)(x, y \in Z)\}$ is the equivalence relation on S corresponding to Y.*

Table 3.1. *Axioms: Rules on Equivalence Relations. For notation see Definitions* 2.7 (*on page* 15) *and* 3.6.

***Cst*$_1$: 1-Constant** $(\forall x, y \in S)(\forall P \in \mathcal{E}^k)(xCPy)$

***Dct*: Dictatorship** $(\exists j \in K)(\forall P \in \mathcal{E}^k)(E_j \subseteq CP)$

***Ind*: Independence** $(\forall X \subseteq S)(\forall P, P' \in \mathcal{E}^k)(P|_X = P'|_X \Longrightarrow CP|_X = CP'|_X)$

***Olg*: Oligarchy** $(\exists V \subseteq K)(\forall P \in \mathcal{E}^k)(CP = \cap_{i \in V} E_i)$

***PO*: Pareto Optimality** $(\forall x, y \in S)(\forall P \in \mathcal{E}^k)(xE_K y \Longrightarrow xCPy)$

***Prj*: Projection** $(\exists j \in K)(\forall P \in \mathcal{E}^k)(CP = E_j)$

***Sym*: Symmetry** $(\forall P \in \mathcal{E}^k)(\forall \text{ permutations } \sigma \text{ of } K)(CP = C(E_{\sigma(1)}, \ldots, E_{\sigma(k)}))$

With this correspondence it is easy to pass between an equivalence relation on S and its natural partition of S. Mirkin's insight was to investigate the problem of aggregating equivalence relations as a way to attack the corresponding problem of aggregating partitions of a set. Thus for the set $\mathcal{E}$ of all equivalence relations on S, consider consensus rules $C : \mathcal{E}^k \longrightarrow \mathcal{E}$ that are subject to Convention 2.4 on page 13. We simplify notation when no confusion arises.

Definition 3.6. *For all* $x, y \in S$ *and* $P = (E_1, \ldots, E_k) \in \mathcal{E}^k$,

shorten	*to*
$C(P)$	CP
$(\forall i \in I \subseteq K)(xE_i y)$	$xE_I y$.

Consider what axioms would be suitable to describe consensus rules on equivalence relations. In view of Arrow's success with weak orders, it is natural (here and in later sections) to formulate axioms of independence, Pareto optimality, and dictatorship. In fact with two exceptions the axioms in Table 3.1 are simple restatements of corresponding axioms for weak orders. The projection (*Prj*) axiom restricts the dictatorial concept to strong dictators who prevent all other individuals from contributing to the consensus result. The projection and constant axioms identify consensus rules that restrict the abilities of most individuals to influence an election's result. The oligarchy (*Olg*) axiom generalizes the dictatorial concept to forms of consensus in which the ruling power belongs to a set of individuals.

Definition 3.7. *The method of* rule by oligarchy $V \subseteq K$ *is the consensus rule* $C_V : \mathcal{E}^k \longrightarrow \mathcal{E}$ *such that*

$$(\forall P \in \mathcal{E}^k)(C_V(P) = C_V P = \cap_{i \in V} E_i).$$

Rule by unanimity *or* strict consensus *is the rule* $Str : \mathcal{E}^k \longrightarrow \mathcal{E}$ *such that*

$$(\forall P \in \mathcal{E}^k)(Str(P) = Str\,P = \textstyle\bigcap_{i \in K} E_i).$$

An oligarchic rule C_V is 1-constant if and only if $V = \emptyset$, it is dictatorial if and only if $V = \{i\}$ for $i \in K$, and it is strict if and only if $V = K$.

3.1.1 Mirkin's Theorem

Mirkin's impossibility theorem for partitions of a set is a characterization of oligarchies. Its import is that nonoligarchies, which are desirable, are impossible in the presence of other desirable properties.

Theorem 3.8 (Mirkin). [282, p. 446]. *Let C be a consensus rule on $\mathcal{E}$. C satisfies Olg if and only if C satisfies Ind and PO.*

When a nonconstant rule C satisfies *Sym*, CP is determined only by the individual relations in P and not by the way the individuals (or subscripts) are associated with those relations. Since $\neg Cst_1$ and *Sym* thus prevent oligarchies with $V \subset K$, Theorem 3.8 yields a characterization of the strict consensus rule.

Corollary 3.9. *Let C be a consensus rule on $\mathcal{E}$. $C = Str$ if and only if C satisfies $\neg Cst_1$, Ind, PO, and Sym.*

For symmetric relations, decisiveness and inverse decisiveness are identical concepts; but whereas decisive sets serve to include ordered pairs in a consensus result, blocking sets serve to exclude them.

Definition 3.10. *Given a consensus rule $C : \mathcal{E}^k \longrightarrow \mathcal{E}$, let $I \subseteq K$ and $x, y \in S$ with $x \neq y$. I is called* decisive for xy, *a condition denoted by U^I_{xy}, if every $P = (E_1, \ldots, E_k) \in \mathcal{E}^k$ having xE_Iy then implies $xCPy$. I is called* decisive *if it is decisive for all $x, y \in S$ with $x \neq y$. U_C is the set of all decisive sets. I is called* blocking for xy, *a condition denoted by B^I_{xy}, if every $P \in \mathcal{E}^k$ having $(\forall i \in I)(\neg xE_iy)$ then implies $\neg xCPy$. I is called* blocking *if it is blocking for all $x, y \in S$ with $x \neq y$. B_C is the set of all blocking sets.*

Lemma 3.11 establishes a requirement of invariant decisiveness.

Lemma 3.11. [283, p. 129]. *Let C be a consensus rule on $\mathcal{E}$ that satisfies Ind and PO. If $I \subseteq K$, then*

$$(\exists ab \in S^2)(U^I_{ab}) \Longleftrightarrow I \in U_C.$$

Proof. Let U^I_{ab} for distinct $a, b \in S$. We first prove that $U^I_{ab} \Longrightarrow U^I_{ax}$ for all $x \neq a$. If $x = b$, U^I_{ax} is true by hypothesis. If $x \neq b$, then using *Ind*, let $P = (E_1, \ldots, E_k) \in \mathcal{E}^k$ be such that

$$(\forall i \in I)(ab \in E_i \wedge ax \in E_i \wedge bx \in E_i) \text{ and}$$
$$(\forall i \in K \setminus I)(ab \notin E_i \wedge ax \notin E_i \wedge bx \in E_i).$$

Then $U_{ab}^I \implies aCPb$ and $PO \implies bCPx$, so $aCPx$ by transitivity, whence U_{ax}^I since aE_ix for $i \in I$. Similarly $U_{ab}^I \implies U_{xb}^I$ for all $x \neq b$. By Lemma 2.12 on page 17, $U_{ab}^I \implies U_{xy}^I$ for all $xy \in S^2$ with $x \neq y$; thus $I \in U_C$. The converse is trivial. □

Lemma 3.12 establishes a requirement of intersection invariance.

Lemma 3.12. [283, p. 129]. *Let C be a consensus rule on $\mathcal{E}$ that satisfies Ind and PO. If $I, J \subseteq K$, then*

$$I \cap J \in U_C \iff I, J \in U_C.$$

Proof. Let $I, J \in U_C$ and for each distinct $x, y, z \in S$ use *Ind* to consider some $P = (E_1, \ldots, E_k) \in \mathcal{E}^k$ having

$$(\forall i \in I \cap J)(xy \in E_i \wedge yz \in E_i \wedge xz \in E_i),$$
$$(\forall i \in I \setminus J)(xy \in E_i \wedge yz \notin E_i \wedge xz \notin E_i),$$
$$(\forall i \in J \setminus I)(xy \notin E_i \wedge yz \in E_i \wedge xz \notin E_i), \text{ and}$$
$$(\forall i \notin I \cup J)(xy \in E_i \wedge yz \notin E_i \wedge xz \notin E_i).$$

Since $I \in U_C$ and xE_Iy, then $xCPy$. Since $J \in U_C$ and yE_Jz, then $yCPz$. Thus $xCPz$ by transitivity, so that $U_{xz}^{I \cap J}$, whence $I \cap J \in U_C$ by Lemma 3.11. The converse follows by the definition of decisive set. □

In fact U_C is a filter in interesting cases.

Lemma 3.13. *If consensus rule $C : \mathcal{E}^k \longrightarrow \mathcal{E}$ satisfies Ind, PO, and $\neg Cst_1$, then U_C is a filter, i.e.,*

$$\emptyset \notin U_C \wedge K \in U_C, \tag{3.1a}$$
$$(\forall I \in U_C)(\forall J \subseteq K)(I \subseteq J \implies J \in U_C) \text{ and} \tag{3.1b}$$
$$(\forall I, J \in U_C)(I \cap J \in U_C). \tag{3.1c}$$

Proof. (3.1a) holds since $\neg Cst_1 \implies \emptyset \notin U_C$ and $PO \implies K \in U_C$. (3.1b) follows by the definition of decisive set. Lemma 3.12 establishes (3.1c). □

If $\{i\} \in U_C$, then i is a dictator and thus influences the result of an election: by voting for $xy \in S^2$, i ensures that xy is in the election's result. If $\{i\} \in B_C$, then i influences an election's result in another way: by not voting for xy, i ensures that xy is not in the election's result. Such individuals are determined by *minimally decisive* sets, i.e., by decisive sets that properly contain no other decisive sets.

Lemma 3.14. [283, p. 130]. *Let C be a consensus rule on $\mathcal{E}$ that satisfies Ind. If $V \in U_C$ is minimally decisive, then $(\forall j \in V)(\{j\} \in B_C)$.*

Proof. If $V = \emptyset$, the result holds trivially, so fix $j \in V$ for $V \in U_C$, and for each distinct $x, y, z \in S$ use *Ind* to consider some $P = (E_1, \ldots, E_k) \in \mathcal{E}^k$ having

$$xy \notin E_j \wedge yz \in E_j \wedge xz \notin E_j,$$
$$(\forall i \in V \setminus \{j\})(xy \in E_i \wedge yz \in E_i \wedge xz \in E_i), \text{ and}$$
$$(\forall i \in K \setminus V)(xy \in E_i \wedge yz \notin E_i \wedge xz \notin E_i).$$

Since $V \in U_C$, then $yCPz$. Since V is minimally decisive, then $V \setminus \{j\} \notin U_C$, so that $\neg xCPz$. If $xCPy$, then $xCPz$ by transitivity, a contradiction; thus $\neg xCPy$, so that $B_{xy}^{\{j\}}$, whence $\{j\} \in B_C$ since x and y were unconstrained. □

With Lemmas 3.13 and 3.14 we obtain this proof.

***Proof of Mirkin's Theorem* 3.8 on page 30.** Let C be a consensus rule on $\mathcal{E}$. If C satisfies *Olg*, then clearly it satisfies *Ind* and *PO*. For the converse let C satisfy *Ind* and *PO*. If C satisfies Cst_1, then it satisfies *Olg* for $V = \emptyset$, so let C satisfy $\neg Cst_1$. Since U_C is nonempty by (3.1a) and finite, use (3.1c) to find the minimally decisive sets for C. Only one exists: if distinct $I, J \in U_C$ were both minimal, then $I \cap J \in U_C$ by (3.1c) and $|I \cap J| < \min\{|I|, |J|\}$, a contradiction of minimal decisiveness. Let $V \in U_C$ be that minimally decisive set. Since $V \neq \emptyset$ by (3.1a), then $\bigcap_{i \in V} E_i \subseteq CP$. By Lemma 3.14, $(\forall i \in V)(\forall x, y \in S)(\neg xE_iy \Longrightarrow \neg xCPy)$, so that $CP \subseteq \bigcap_{i \in V} E_i$. Thus $CP = \bigcap_{i \in V} E_i$, whence C satisfies *Olg* for this V. □

Mirkin's theorem yields characterizations of dictatorships, which show that always such dictatorships are strong.

Corollary 3.15. *Let C be a consensus rule on $\mathcal{E}$ that satisfies Ind and PO. These are equivalent:* (i) *C satisfies Dct.* (ii) *C satisfies Prj.* (iii) *U_C is an ultrafilter.*

Proof. Clearly (i) and (ii) are equivalent. Let $i \in K$ be a dictator so that $\{i\} \in U_C$. Since U_C is a filter by Lemma 3.13, we have only to establish

$$(\forall I \subseteq K)(K \setminus I \notin U_C \Longrightarrow I \in U_C); \tag{3.2}$$

but if $I \subseteq K \wedge K \setminus I \notin U_C$, then $\{i\} \subseteq I$; thus $I \in U_C$ by (3.1b), whence (3.2) holds. Let U_C be an ultrafilter and let $K_i = K \setminus \{i\}$ for all $i \in K$. If i is a dictator for some $i \in K \setminus \{k\}$, we are done; otherwise $K_i \in U_C$ by (3.2) for all $i \in K \setminus \{k\}$, so that $\bigcap_{i=1}^{k-1} K_i = \{k\} \in U_C$ by (3.1c), whence k is a dictator. □

This proof makes clear that, for finite K, ultrafilters are precisely the families of sets that contain some fixed element $j \in K$, where j identifies a dictator.

Open Problem 3.16. (M. F. Janowitz.) When *Ind* and *PO* hold for consensus rules on weak orders, then dictators are weak; when they hold for rules on equivalence relations, then dictators are strong. How does the type of relation determine whether dictators are weak or strong? ■

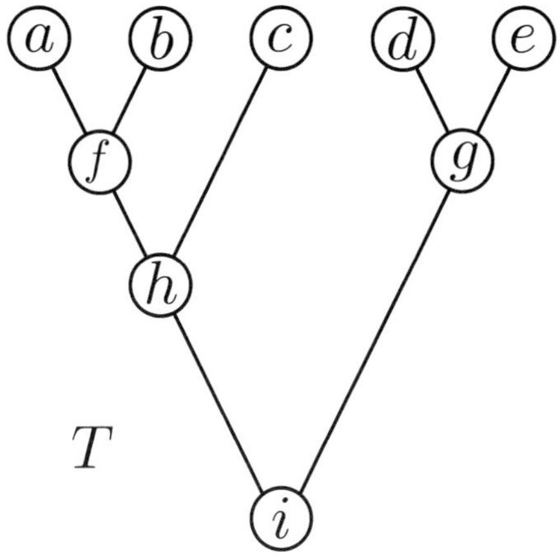

Figure 3.1. *Representing Ancestor-Descendant Relations. Tree T is rooted at vertex i, from which the other vertices are descended. It depicts the phylogenetic relationships among multicellular organisms whose genomes have been sequenced* [64, p. 61]. *a: human (mammal), b: fruit fly (arthropod), c:* Caenorhabditis elegans *(nematode), d:* Arabidopsis *(dicotyledonous plant), e: rice (monocotyledonous plant).*

3.2 Tree Quasi-orders

Let S be a set of elements called *evolutionary units* and consider the ancestor-descendant relationships on S [158]. Biologists can depict these relations by one or more treelike structures such as T in Figure 3.1. By making the strong simplifying assumption that there are no unobserved ancestors, they use S to label every vertex of T. One vertex, $s_0 \in S$, is named the *root*. If T has a path from s_0 to s_j consisting of the edge sequence $\{s_0, s_1\}, \{s_1, s_2\}, \ldots, \{s_{j-2}, s_{j-1}\}, \{s_{j-1}, s_j\}$, then s_{j-1} is the immediate ancestor of s_j, s_{j-2} is the immediate ancestor of s_{j-1}, etc. As a minimal assumption McMorris and Neumann [267] model such an evolutionary history by a tree quasi-order (Table 1.3 on page 6) on S.

Example 3.17. Among the 26 tree quasi-orders on $S = abc$, these representatives are depicted in Figure 3.2:

$$\begin{aligned} R_1 &= \{aa, bb, cc\}, & R_5 &= \{aa, bb, cc, ab, ac, bc\}, \\ R_2 &= \{aa, bb, cc, ab\}, & R_6 &= \{aa, bb, cc, ab, ba, ac, bc\}, \\ R_3 &= \{aa, bb, cc, ab, ba\}, & R_7 &= \{aa, bb, cc, ab, ac, bc, cb\}, \\ R_4 &= \{aa, bb, cc, ab, cb\}, & R_8 &= \{aa, bb, cc, ab, ba, ac, ca, bc, cb\}. \end{aligned}$$

Notice in Figure 3.2 on the following page that the graph-theoretic depiction of a tree quasi-order may have more than one rooted component, e.g., T_1–T_3. ∎

Just as a weak order imposes a linear ordering on the classes of a partition of S, so a tree quasi-order $R \subseteq S^2$ imposes a treelike ancestral ordering on a partition's classes: for all $x, y \in S$, xRy if and only if y is ancestral to x. For all $x \in S$, reflexivity asserts that x is

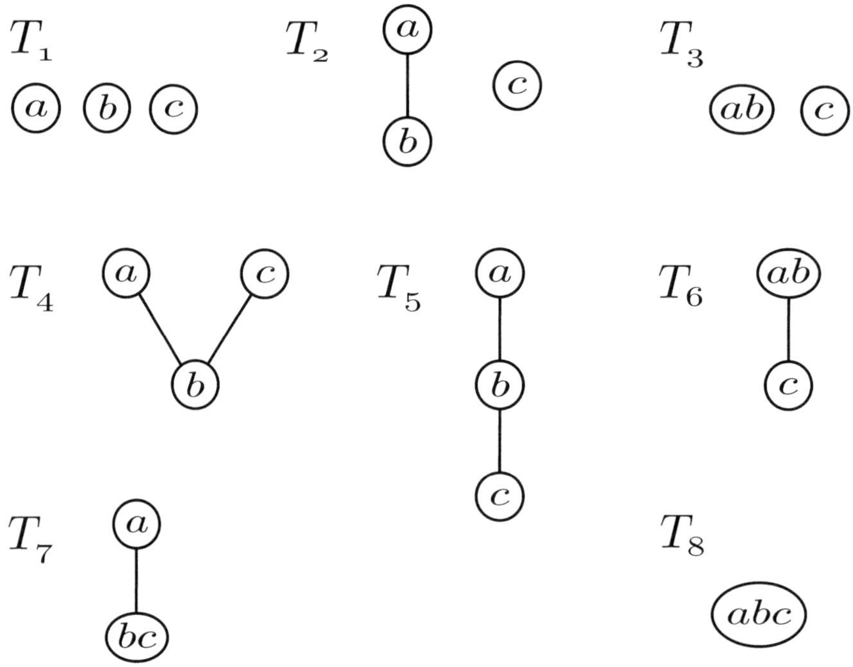

Figure 3.2. *Tree Quasi-orders on Three Evolutionary Units. Each graph T_i depicts the corresponding tree quasi-order R_i in Example* 3.17. *The lowest vertex of a connected component is ancestral to other vertices in that component.*

trivially ancestral to itself. For all $x, y, z \in S$ such that z is ancestral to y and y is ancestral to x, transitivity asserts that z is ancestral to x. For all $x, y, z \in S$ such that y and z are ancestors of x, the tree condition asserts that one of $\{y, z\}$ is ancestral to the other. R defines strict ancestry via its asymmetric part $R^\star$, where $(\forall x, y \in S)(xR^\star y \iff xRy \wedge \neg yRx)$.

Example 3.18. For tree T in Figure 3.1, let $S = abcdefghi$ and consider relation

$$R = \{aa, bb, cc, dd, ee, ff, gg, hh, ii, af, bf, ch, dg, eg, fh, gi, hi, ah, ai, bh, bi, ci, di, ei\}.$$

The first nine pairs represent the vertices of T, the next eight represent its edges, and the last seven follow by transitivity. Readers may verify that R satisfies the tree condition inasmuch as, e.g., *fRh* as required by the occurrences of *aRf* and *aRh*. ■

Let $\mathcal{T}$ be the set of all tree quasi-orders on S, and consider consensus rules $C : \mathcal{T}^k \longrightarrow \mathcal{T}$ that are subject to Convention 2.4 on page 13. We simplify notation when no confusion arises.

Table 3.2. *Axioms: Rules on Tree Quasi-orders. If $R \in \mathcal{T}$, then $R^\star$ denotes strict ancestral relationship; for other notation see Definitions* 2.7 *on page* 15 *and* 3.19.

***Dct*: Dictatorship** $(\exists j \in K)(\forall x, y \in S)(\forall P \in \mathcal{T}^k)(xR_j^\star y \Longrightarrow xCP^\star y)$

***Ind*: Independence** $(\forall X \subseteq S)(\forall P, P' \in \mathcal{T}^k)(P|_X = P'|_X \Longrightarrow CP|_X = CP'|_X)$

***PO*: Pareto Optimality** $(\forall x, y \in S)(\forall P \in \mathcal{T}^k)(xR_K^\star y \Longrightarrow xCP^\star y)$

Definition 3.19. *For all $x, y \in S$ and $P = (R_1, \ldots, R_k) \in \mathcal{T}^k$,*

shorten	*to*
$C(P)$	CP
$(\forall i \in I \subseteq K)(xR_i^\star y)$	$xR_I^\star y$.

Consider what axioms would be suitable to describe consensus rules on tree quasi-orders. In view of Arrow's success with weak orders, again it is natural to formulate axioms of independence, Pareto optimality, and dictatorship. In fact, the formulations in Table 3.2 of axioms for tree quasi-orders are simple restatements of corresponding axioms for weak orders in Table 2.3 on page 14.

3.2.1 McMorris and Neumann's Theorem

The import of McMorris and Neumann's impossibility theorem for tree quasi-orders is that nondictatorships, which are desirable, are impossible in the presence of other equally desirable properties.

Theorem 3.20 (McMorris and Neumann). [267, p. 132]. *Let C be a consensus rule on $\mathcal{T}$. C satisfies Dct if it satisfies Ind and PO.*

This result follows from familiar and new concepts of decisiveness.

Definition 3.21. *Given a consensus rule $C : \mathcal{T}^k \longrightarrow \mathcal{T}$, let $I \subseteq K$ and $x, y \in S$ with $x \neq y$. I is called* almost decisive for xy, *a condition denoted by $\widehat{U}^I_{xy}$, if every $P = (R_1, \ldots, R_k) \in \mathcal{T}^k$ having $xR_I^\star y$ and $yR_{K\setminus I}^\star x$ then implies $xCP^\star y$. I is called* almost decisive *if it is almost decisive for all $x, y \in S$ with $x \neq y$. $\widehat{U}_C$ is the set of all almost decisive sets. I is called* decisive for xy, *a condition denoted by U^I_{xy}, if every $P = (R_1, \ldots, R_k) \in \mathcal{T}^k$ having $xR_I^\star y$ then implies $xCP^\star y$. I is called* decisive *if it is decisive for all $x, y \in S$ with $x \neq y$. U_C is the set of all decisive sets.*

Lemma 3.22 establishes a requirement of invariant decisiveness.

Lemma 3.22. [267, p. 133]. *Let C be a consensus rule on $\mathcal{T}$ that satisfies Ind and PO. If $I \subseteq K$, then*

$$(\exists ab \in S^2)(\widehat{U}^I_{ab}) \iff I \in U_C.$$

Proof. Let $\widehat{U}^I_{xy}$ for $xy \in S^2$. We first prove that $\widehat{U}^I_{xy} \implies U^I_{zy}$ for all $z \neq y$. Suppose $P \in \mathcal{T}^k$ satisfies $zR^\star_I y$. Assume $z \neq x$ and construct $P' \in \mathcal{T}^k$ to have $zR'^\star_I x$, $xR'^\star_I y$, $yR'^\star_{K\setminus I}x$, $zR'^\star_{K\setminus I}x$, and $R'_j|_{\{y,z\}} = R_j|_{\{y,z\}}$ for all $j \in K \setminus I$. Since $PO \implies zCP'^\star x$ and $\widehat{U}^I_{xy} \implies xCP'^\star y$, then $zCP'^\star y$ by transitivity. Using P, P' and $X = \{y, z\}$, *Ind* $\implies zCP^\star y$, whence U^I_{zy} for all $z \neq x, y$. U^I_{xy} now follows by interchanging x and z in the preceding argument, whence U^I_{zy} for all $z \neq y$. Similarly $\widehat{U}^I_{xy} \implies U^I_{xz}$ for all $z \neq x$. By a slight variant of Lemma 2.12 on page 17, $\widehat{U}^I_{xy} \implies U^I_{zw}$ for all $zw \in S^2$ with $z \neq w$, whence $I \in U_C$. The converse is trivial. □

Lemma 3.23 ensures that a form of completeness (Table 1.2 on page 6) in $P \in \mathcal{T}^k$ induces a corresponding completeness in CP.

Lemma 3.23. [267, p. 133]. *Let C be a consensus rule on $\mathcal{T}$ that satisfies Ind and PO. If $P = (R_1, \ldots, R_k) \in \mathcal{T}^k$ and $y, z \in S$, then*

$$(\forall i \in K)(yR_i z \vee zR_i y) \implies yCPz \vee zCPy.$$

Proof. Choose $x \in S \setminus \{y, z\}$ and construct $P' \in \mathcal{T}^k$ to have $xR'^\star_K y$, $xR'^\star_K z$, and $(\forall i \in K)(R'_i|_{\{y,z\}} = R_i|_{\{y,z\}})$. *PO* implies that $xCP'^\star y \wedge xCP'^\star z$. Since CP satisfies the tree condition, then $yCP'z \vee zCP'y$; the result follows by using *Ind* with P, P' and $X = \{y, z\}$. □

With Lemmas 3.22 and 3.23 we obtain this proof.

***Proof of McMorris and Neumann's Theorem* 3.20.** Let rule $C : \mathcal{T}^k \longrightarrow \mathcal{T}$ satisfy *Ind* and *PO*. The main task is to prove that $(\exists j \in K)(\{j\} \in \widehat{U}_C)$; then Lemma 3.22 implies that $(\exists j \in K)(\{j\} \in U_C)$, i.e., that a dictator exists, and we are done. To begin, since *PO* implies that $K \in \widehat{U}_C$, let $I \subseteq K$ be an almost decisive set of minimum cardinality. If $|I| > 1$, then a contradiction arises in the following way. Choose $j \in I$ arbitrarily and for pairwise distinct $x, y, z \in S$ construct a profile $P \in \mathcal{T}^k$ to have $xR^\star_{\{j\}}y$, $yR^\star_{\{j\}}z$, $zR^\star_{I\setminus\{j\}}x$, $xR^\star_{I\setminus\{j\}}y$, $yR^\star_{K\setminus I}z$, and $zR^\star_{K\setminus I}x$, so that by transitivity P has $xR^\star_{\{j\}}z$, $zR^\star_{I\setminus\{j\}}y$, and $yR^\star_{K\setminus I}x$. Since $\widehat{U}^I_{xy}$, then $xCP^\star y$. For $y, z \in S$, since P satisfies the hypotheses of Lemma 3.23, we have $yCPz$ or $zCPy$ (or both). If $zCP^\star y$, then $\widehat{U}^{I\setminus\{j\}}_{zy}$, which is false; thus $yCPz$ or $yCP^\star z$, so $yCPz$ holds in any case. Since $xCP^\star y$ and $yCPz$, then $xCP^\star z$ by transitivity. But now we may apply *Ind* with $X = \{x, z\}$ to conclude that $\widehat{U}^{\{j\}}_{xz}$, which is false. Thus assuming $|I| > 1$ is contradictory, so $(\exists j \in K)(\{j\} \in \widehat{U}_C)$ as was to be proved. □

Open Problem 3.24. Characterize consensus rules on tree quasi-orders that satisfy *Ind*. ■

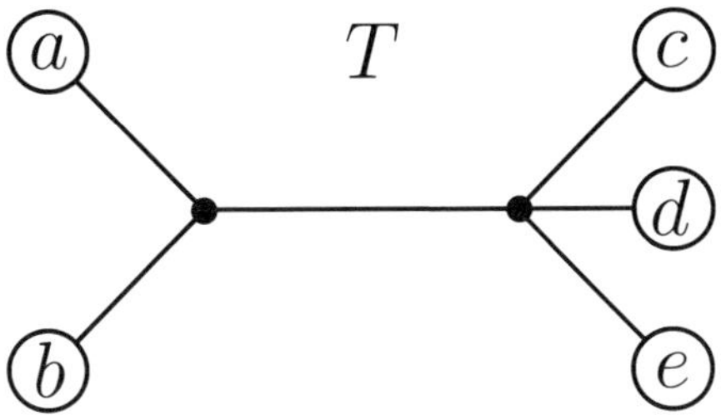

Figure 3.3. *A Phylogeny*

3.3 Phylogenies

We have obtained impossibility results for weak orders (Theorem 2.9 on page 17) and tree quasi-orders (Theorem 3.20 on page 35). These objects exhibit a natural order, e.g., preference of one class of a partition to another; descent of one evolutionary unit from another. Would impossibility results hold if such a natural order is absent? Consider the case of unrooted phylogenies on a set S.

Definition 3.25. *A* phylogeny *on* $S = S_n$ *is a graph-theoretic tree with no vertices of degree* 2 *and exactly* n *vertices of degree* 1 *(the leaves), each labeled by a distinct element of* S *(Figure* 3.3*). Let* $\mathcal{P} = \mathcal{P}_n$ *be the set of all phylogenies on* $S = S_n$.

Convention 3.26. *In the biological literature a phylogeny may be either unrooted or rooted. In this book, unless otherwise qualified, a phylogeny will be unrooted, and the term hierarchy will denote the rooted phylogeny of the biologist.*

Every phylogeny is representable by a quaternary relation on S whose 4-tuples are called resolved quartets.

Definition 3.27. *For all* $T \in \mathcal{P}$ *and for all distinct* $w, x, y, z \in S$*, let* $wx|yz$ *denote the configuration in* T *where the path between* w *and* x *has no vertices of the path between* y *and* z*, and let* $wxyz$ *denote the one where for each partition of* $\{w, x, y, z\}$ *into two pairs, the corresponding paths intersect at a unique interior vertex of* T*. The configurations* $wx|yz$*,* $wy|xz$*,* $wz|xy$*, and* $wxyz$ *(Figure* 3.4*) are called* quartets*; the first three are called* resolved *while* $wxyz$ *remains* unresolved.

Colonius and Schulze [136] showed that every $T \in \mathcal{P}$ is uniquely determined by specifying which quartet is in T for each four-element subset of S; so we may assume that every $T \in \mathcal{P}$ is represented by the set $q(T)$ of its quartets. We will shorten $q(T)$ to T where no confusion arises.

Example 3.28. Since $S = abcde$ has five four-element subsets, the phylogeny in Figure 3.3 is represented by

$$q(T) = T = \{ab|cd, ab|ce, ab|de, acde, bcde\}. \quad \blacksquare$$

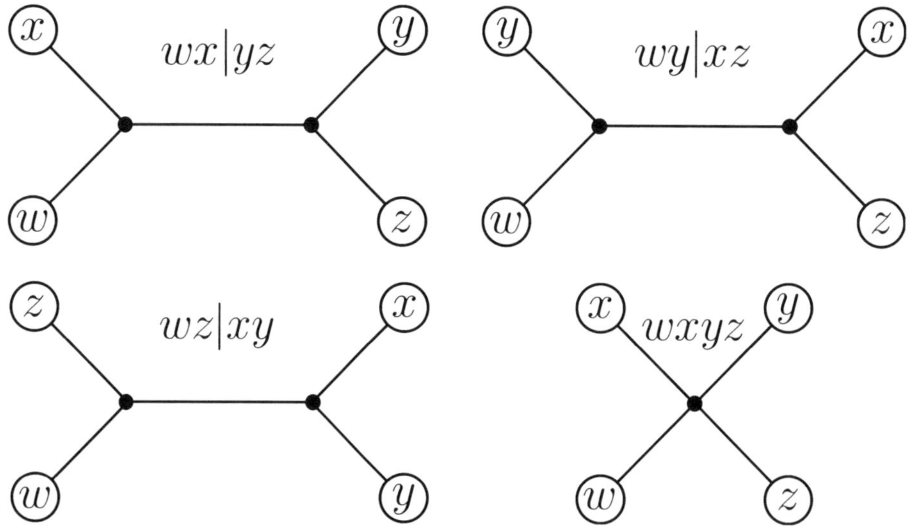

Figure 3.4. *Quartets for Representing Phylogenies*

The restriction concept of binary relations (Definition 2.7 on page 15) extends naturally to phylogenies.

Definition 3.29. *For all $X \subseteq S$ and $T \in \mathcal{P}$, let the set of quartets of T made up entirely with elements of X be called the* restriction $T|_X$ *of T to X. For each $P = (T_1, \ldots, T_k) \in \mathcal{P}^k$, let $P|_X = (T_1|_X, \ldots, T_k|_X)$.*

Consider consensus rules $C : \mathcal{P}^k \longrightarrow \mathcal{P}$ that are subject to this convention.

Convention 3.30. *To any (multi)consensus rule C with domain $\mathcal{P}^k$ is associated a set $S = S_n$ of n leaf labels on which the phylogenies are defined. In this context, unless specifically stated otherwise, S is finite with $|S| = n \geq 5$.*

We simplify notation for consensus rules on $\mathcal{P}$ when no confusion arises.

Definition 3.31. *For all $w, x, y, z \in S$ and $P = (T_1, \ldots, T_k) \in \mathcal{P}^k$,*

shorten	*to*
$C(P)$	CP
$q(T_i)$	T_i
$wx\|yz \in q(T_i)$	wxT_iyz
$wxyz \in q(T_i)$	$wxyzT_i$
$(\forall i \in I \subseteq K)(wxT_iyz)$	wxT_Iyz
$(\forall i \in I \subseteq K)(wxyzT_i)$	$wxyzT_I$.

Table 3.3. *Axioms: Rules on Phylogenies* I. *For notation see Definitions* 3.29 *and* 3.31 *on the preceding page.*

***Cst*: Constant** $(\exists T \in \mathcal{P})(\forall P \in \mathcal{P}^k)(CP = T)$

***Dct*: Dictatorship** $(\exists j \in K)(\forall w, x, y, z \in S)(\forall P \in \mathcal{P}^k)(wxT_jyz \Longrightarrow wxCPyz)$

***Ind*: Independence** $(\forall X \subseteq S)(\forall P, P' \in \mathcal{P}^k)(P|_X = P'|_X \Longrightarrow CP|_X = CP'|_X)$

***PO*: Pareto Optimality** $(\forall w, x, y, z \in S)(\forall P \in \mathcal{P}^k)(wxT_Kyz \Longrightarrow wxCPyz)$

***Prj*: Projection** $(\exists j \in K)(\forall P \in \mathcal{P}^k)(CP = T_j)$

Consider what axioms would be suitable to describe consensus rules on phylogenies. In view of Arrow's success with weak orders, again it is natural to formulate axioms of independence, Pareto optimality, and dictatorship. In fact, the formulations in Table 3.3 of axioms for phylogenies are simple restatements of corresponding axioms for equivalence relations in Table 3.1 on page 29.

3.3.1 McMorris and Powers's Theorems

McMorris and Powers establish for phylogenies a characterization of dictatorships. The import of their results is that nondictatorships, which are desirable, are impossible in the presence of other equally desirable properties.

Theorem 3.32 (McMorris and Powers). [269, p. 49]. *Let C be a consensus rule on $\mathcal{P}$. C satisfies Dct if it satisfies Ind and PO.*

Our proof uses Sen's strategy (Table 2.4 on page 19), where $m = 4$ and steps 1–3 correspond to Definition 3.33 and Lemmas 3.34 and 3.35, which follow. Familiar concepts of decisiveness are relevant.

Definition 3.33. *Given a consensus rule $C : \mathcal{P}^k \longrightarrow \mathcal{P}$, let $I \subseteq K$ and let $wx|yz$ be a quartet with $\{w, x, y, z\} \subseteq S$. I is called* almost decisive for *$wx|yz$, a condition denoted by $\widehat{U}^I_{wx|yz}$, if every $P = (T_1, \ldots, T_k) \in \mathcal{P}^k$ having wxT_Iyz and $wxyzT_{K\setminus I}$ then implies $wxCPyz$. I is called* almost decisive *if it is almost decisive for all resolved quartets. $\widehat{U}_C$ is the set of all almost decisive sets. I is called* decisive for *$wx|yz$, a condition denoted by $U^I_{wx|yz}$, if every $P = (T_1, \ldots, T_k) \in \mathcal{P}^k$ having wxT_Iyz then implies $wxCPyz$. I is called* decisive *if it is decisive for all resolved quartets. U_C is the set of all decisive sets.*

Lemma 3.34 establishes requirements of invariant decisiveness.

Lemma 3.34. [269, p. 50]. *Let C be a consensus rule on $\mathcal{P}$ that satisfies Ind and PO. If $I \subseteq K$, then*

$$(\exists abcd \in S^4)(\widehat{U}^I_{ab|cd}) \Longleftrightarrow I \in \widehat{U}_C \text{ and}$$
$$(\exists abcd \in S^4)(U^I_{ab|cd}) \Longleftrightarrow I \in U_C.$$

Proof. For almost decisiveness let $\widehat{U}^I_{ab|cd}$, and since $|S| \geq 5$ let $v \in S$ be such that $v \notin X = abcd$; then we will show that $\widehat{U}^I_{bv|cd}$. Construct $P \in \mathcal{P}^k$ to have $\{ab|cd, ab|cv, ab|dv, av|cd, bv|cd\} \subseteq T_I$ and $\{abcd, av|bc, av|bd, av|cd, bcdv\} \subseteq T_{K\setminus I}$, P being otherwise unconstrained. Since $\widehat{U}^I_{ab|cd} \Longrightarrow abCPcd$ and $PO \Longrightarrow avCPcd$, it is easily shown that $bvCPcd$, whence $Ind \Longrightarrow \widehat{U}^I_{bv|cd}$. By trivial variants of this argument we obtain $\widehat{U}^I_{wx|yz}$ for each $wx|yz$ other than $ab|cd$, whence $I \in \widehat{U}_C$. The converse is trivial.

For decisiveness let $(\exists abcd \in S^4)(U^I_{ab|cd})$, then clearly $\widehat{U}^I_{ab|cd}$, so that $I \in \widehat{U}_C$ by the previous argument. Now let $P \in \mathcal{P}^k$ and $X = wxyz \subseteq S$; we will show that $wxCPyz$ if $\{wx|yz\} \subseteq T_I$, whence $I \in U_C$. Since $|S| \geq 5$, let $v \notin X$ and construct $P' \in \mathcal{P}^k$ to have $\{wx|yz, wx|vy, wx|vz, vwyz, vxyz\} \subseteq T'_I$, $\{vwxy, vwxz\} \subseteq T'_{K\setminus I}$, and $P'|_X = P|_X$, P' being otherwise unconstrained. Since $I \in \widehat{U}_C$, then $\widehat{U}^I_{wx|vy}$ and $\widehat{U}^I_{wx|vz}$, so $\{wx|vy, wx|vz\} \subseteq CP'$, whence it is easily shown that $wxCP'yz$. But $P'|_X = P|_X$, so *Ind* implies $wxCPyz$ as required. The converse is trivial. □

Lemma 3.35 establishes an invariance requirement of equivalent subsets.

Lemma 3.35. *Let C be a consensus rule on $\mathcal{P}$ that satisfies Ind and PO. If I and J satisfy $I \subseteq J \subseteq K$, then*

$$J \setminus I \notin U_C \Longrightarrow [I \in U_C \Longleftrightarrow J \in U_C].$$

Proof. Assume $J \setminus I \notin U_C$. Clearly $I \in U_C \Longrightarrow J \in U_C$, so instead let $J \in U_C$. Construct $P \in \mathcal{P}^k$ to have $\{ab|cd, ae|cd\} \subseteq T_I$ and $\{ab|cd, ab|ce\} \subseteq T_{J\setminus I}$, P being otherwise unconstrained. Since $J \in U_C$, then $abCPcd$. If $ab|ce \in CP$, then $J \setminus I \in U_C$, a contradiction, so $ab|ce \notin CP$. If $ab|cd \in CP$ and $ab|de \in CP$, then $ab|ce \in CP$, a contradiction, so $ab|de \notin CP$. But $ab|cd \in CP$ and $ab|de \notin CP$ imply that $ae|cd \in CP$, so $U^I_{ae|cd}$, whence $I \in U_C$ by Lemma 3.34. □

With Lemma 3.35 we obtain this proof.

***Proof of McMorris and Powers's Theorem* 3.32.** Let rule $C : \mathcal{P}^k \longrightarrow \mathcal{P}$ satisfy *Ind* and *PO*. Since K is decisive (by *PO*) and finite, recursive partitioning by equivalent subsets (using Lemma 3.35) shows that $\{i\} \in U_C$ for some $i \in K$, whence *Dct* holds. □

Theorem 3.32 and Lemma 3.35 easily yield this corollary.

Corollary 3.36. *If consensus rule $C : \mathcal{P}^k \longrightarrow \mathcal{P}$ satisfies Ind and PO, then U_C is an ultrafilter.*

With phylogenies the independence axiom is so strong that it completely excludes individuals other than the dictator from determining the result of an election. In this context the natural analogue of Wilson's theorem is the following theorem.

Theorem 3.37. [269, p. 51]. *Let C be a consensus rule on $\mathcal{P}$. C satisfies Ind if and only if C satisfies either Cst or Prj.*

The reader may consult [269] for McMorris and Powers's proof of this result: they use induction on $|S|$ and find that the basis step is difficult to establish. Theorem 3.37 on the facing page shows us that the strengths of axioms such as *Ind* may vary (unpredictably) from one context to another; it also yields a characterization of dictatorial rules.

Corollary 3.38. [269, p. 54]. *Let C be a consensus rule on $\mathcal{P}$. C satisfies Ind and PO if and only if C satisfies Prj.*

Proof. If C satisfies *Ind* and *PO*, then C is a constant or a projection by Theorem 3.37, but *PO* ensures that C is not a constant, so it is a projection. If C is a projection, then C satisfies *Ind* (by Theorem 3.37) and *PO* (by its definition). □

Open Problem 3.39. Since *Ind* and *PO* appear to be so strong for consensus rules on phylogenies, how might they be weakened so as to obtain characterizations of rules on phylogenies that are perhaps more relevant than projections? ■

3.4 Hierarchies

It is not, I am persuaded, that memory is random, brutally indifferent, but rather that it has its own strict hierarchies, which are hidden from us.
— T. Flanagan [169, p. 394]

The quotation appears to suggest that the retrieval of information from human memory is based on its hierarchical representation therein. At least in biological taxonomy it is common to depict relationships by hierarchical representations or other treelike structures [184]. To model such structures let S be a set of $n > 0$ elements. Let $T_r = (V, E)$ be a rooted tree with n leaves such that the root $r \in V$ is a vertex with degree at least 2, every other interior vertex has degree at least 3, and each leaf is labelled with a distinct singleton set $\{x\}$, $x \in S$. Draw T_r on a sheet of paper so that the leaves are at the top and r is at the bottom. Label each interior vertex v of T_r by the union of the leaf labels above v in T_r, so r is labeled by S (Figure 3.5). Let $H = h(T_r) \subseteq 2^S$ be the set of all labels associated with the vertices of T_r. Always $|H| > n$ since $S \in H$ and $\{x\} \in H$ for all $x \in S$. Moreover for all $X, Y \in H$ only three possibilities exist: $X \subseteq Y$, so the path in T_r from r to the vertex labeled by X includes the vertex labeled by Y; $Y \subseteq X$, so the path from r to the vertex labeled by Y includes the vertex labeled by X; $X \cap Y = \emptyset$, so no path from r to any leaf includes both vertices labeled by X and Y. Indeed $h(T_r)$ contains so much information on T_r that T_r can be reconstructed from $h(T_r)$. For this reason the sets $h(T_r)$ are called hierarchies and are studied in their own right.

Definition 3.40. *A* (strong) hierarchy *on S is a set $H \subseteq 2^S$ such that $\emptyset \notin H$, $S \in H$, $(\forall x \in S)(\{x\} \in H)$ and $(\forall X, Y \in H)(X \cap Y \in \{\emptyset, X, Y\})$, this being for hierarchies what the tree condition (Table 1.2 on page 6) is for tree quasi-orders. Any $X \in H$ is called a* cluster *and is* nontrivial *if $1 < |X| < |S|$, so let H^* be the set of all nontrivial*

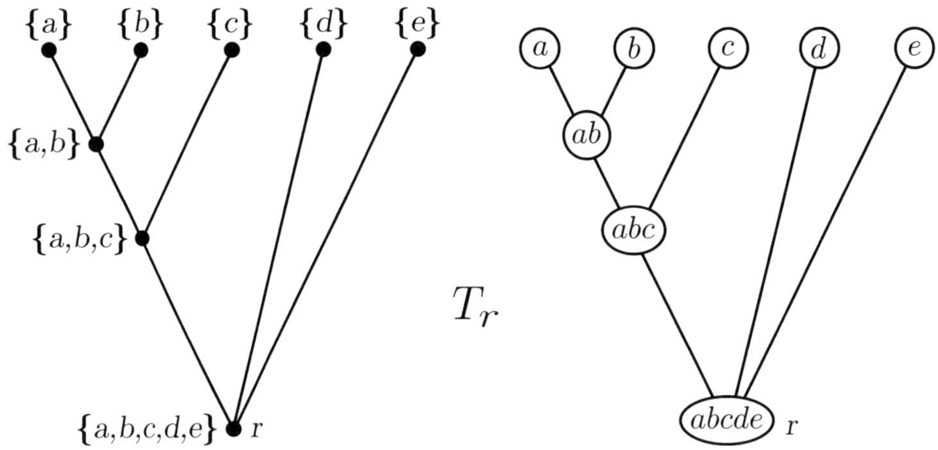

Figure 3.5. *Labeling Vertices of Rooted Trees by Sets of Leaf Labels. As usual we shorten* $\{a\}$ *to* a, $\{a, b\}$ *to* ab, *etc., as shown on the right.*

clusters of H. *Let* $X \in H^*$ *be called* maximal *if* $(\forall Y \in H^*)\,(X \subseteq Y \implies X = Y)$. $H_\emptyset = \{S\} \cup \{\{x\} : x \in S\}$ *is the* null *hierarchy that has no nontrivial clusters. Let* $\mathcal{H} = \mathcal{H}_n$ *be the set of all hierarchies on* $S = S_n$.

The restriction concept has natural interpretations for hierarchies.

Definition 3.41. *For all* $H \in \mathcal{H}$ *and* $\emptyset \neq X \subseteq S$,

$$H|_X = H_\emptyset \cup \{A \cap X : A \in H^* \wedge A \cap X \neq \emptyset\}$$

is the restriction of H to X, *while*

$$H|_X - X = H|_X \setminus \{X\}$$

is the removal restriction of H to X. *For all* $P = (H_1, \ldots, H_k) \in \mathcal{H}^k$, *let* $P|_X = (H_1|_X, \ldots, H_k|_X)$ *and* $P|_X - X = (H_1|_X - X, \ldots, H_k|_X - X)$.

Example 3.42. If $S = abcde$, the tree T_r in Figure 3.5 is representable by the hierarchy $H = h(T_r) = H_\emptyset \cup H^*$, where $H^* = \{ab, abc\}$, whence $H|_{ac} = H_\emptyset \cup \{ac\}$ and $H|_{ac} - ac = H_\emptyset$. ■

A rooted tree is also representable by what is essentially a ternary relation.

Definition 3.43. *For all* T_r *on the leaf set* S *and for all distinct* $x, y, z \in S$, *let* $xy|z$ *denote the configuration in* T_r *where the path between* x *and* y *has no vertices of the path between* z *and* r, *and let* xyz *denote the one where the path between any two leaves in* $\{x, y, z\}$ *has a*

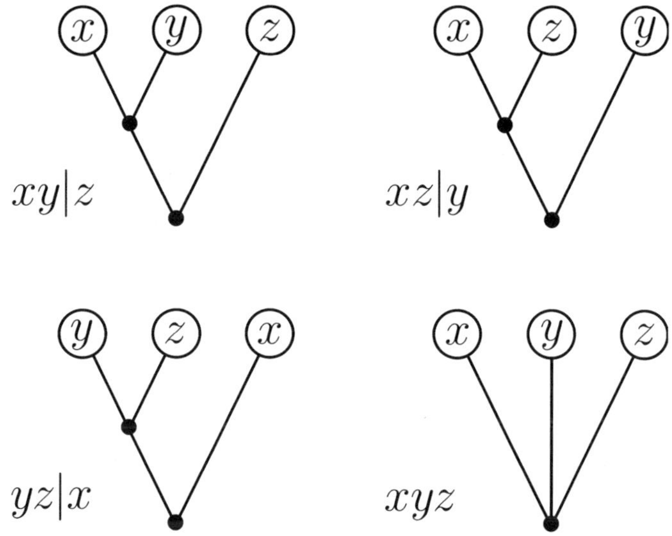

Figure 3.6. *Triads for Representing Hierarchies*

vertex of the path between the third leaf and r. *The configurations* $xy|z$, $xz|y$, $yz|x$, xyz *are called* triads *(Figure 3.6); the first three are called* resolved, *while* xyz *remains* unresolved.

Triads can be defined directly in terms of H:

$$\begin{aligned} xy|z &\iff (\exists Y \in H)(x, y \in Y \wedge z \notin Y), \\ xz|y &\iff (\exists Y \in H)(x, z \in Y \wedge y \notin Y), \\ yz|x &\iff (\exists Y \in H)(y, z \in Y \wedge x \notin Y), \\ xyz &\iff (\forall Y \in H)(|Y \cap \{x, y, z\}| \neq 2). \end{aligned}$$

Colonius and Schulze [136] established that any $H \in \mathcal{H}$ is uniquely determined by specifying which triad occurs in H for each three-element subset of S; so we may assume that every hierarchy H is represented by the set $t(H)$ of its triads.

Example 3.44. If $S = abcde$, then $H = h(T_r) = H_\emptyset \cup \{ab, abc\}$ for the tree T_r in Figure 3.5. Since S has ten three-element subsets, H is representable by the set $t(H) = \{ab|c, ab|d, ab|e, ac|d, ac|e, ade, bc|d, bc|e, bde, cde\}$ of triads. ■

Several of the following developments apply not only to $\mathcal{H}$, the set of hierarchies, but to $\mathcal{W}$ or $\mathcal{W}_c$, the sets of weak (Definition 3.60 on page 48) or closed weak (Definition 3.64 on page 50) hierarchies. Consider consensus rules $C : \mathcal{H}^k \longrightarrow \mathcal{H}$ that are subject to this convention.

Convention 3.45. *To any (multi)consensus rule C with domain $\mathcal{H}^k$ or $\mathcal{W}^k$ is associated a set $S = S_n$ of n leaf labels on which the hierarchies are defined. In this context, unless specifically stated otherwise, S is finite with $|S| = n \geq 5$.*

For example, with the method of majority rule for $\mathcal{H}$, any cluster in more than half of the hierarchies in a profile P will be in the majority rule consensus for P.

Definition 3.46. *The* index γ *of $X \subseteq S$ in $P = (H_1, \ldots, H_k) \in \mathcal{H}^k$ is the proportion of occurrences of X in the hierarchies of P:*

$$\gamma(X, P) = \frac{|\{i \in K : X \in H_i\}|}{k}.$$

Definition 3.47. Margush and McMorris [246]. *The method of* majority rule *for hierarchies is the function $Maj : \mathcal{H}^k \longrightarrow \mathcal{H}$ such that*

$$(\forall P \in \mathcal{H}^k)\left(Maj(P) = Maj P = \left\{X \subseteq S : \gamma(X, P) > \frac{1}{2}\right\}\right).$$

The majority rule on hierarchies is well defined.

Lemma 3.48. [246, p. 242]. $(\forall P \in \mathcal{H}^k)(Maj P \in \mathcal{H})$.

Proof. Clearly $\emptyset \notin Maj P$, $S \in Maj P$, and $(\forall x \in S)(\{x\} \in Maj P)$. If $X, Y \in Maj P$, then since $\gamma(X, P) > \frac{1}{2}$ and $\gamma(Y, P) > \frac{1}{2}$, some $H_i \in P$ has $X, Y \in H_i$, whence $X \cap Y \in \{\emptyset, X, Y\}$. □

We simplify notation for consensus rules on $\mathcal{H}$ when no confusion arises.

Definition 3.49. *For $x, y, z \in S$ and $P = (H_1, \ldots, H_k) \in \mathcal{X}^k$ with $\mathcal{X} \in \{\mathcal{H}, \mathcal{W}, \mathcal{W}_c\}$, where $\mathcal{W}$ and $\mathcal{W}_c$ appear in Definition* 3.60 *on page* 48 *and Definition* 3.64 *on page* 50*,*

shorten	*to*
$C(P)$	CP
$(\forall i \in I \subseteq K)(X \in H_i)$	$X \in H_I$
$\{i \in I \subseteq K : X \in H_i\}$	$I_X(P)$
$t(H_i)$	R_i
$t(CP)$	R
$xy\|z \in t(H_i)$	xyR_iz
$xyz \in t(H_i)$	$xyzR_i$
$(\forall i \in I \subseteq K)(xyR_iz)$	$xyR_Iz.$

If $I = K$, for example, then $K_X(P) = \{i \in K : X \in H_i\}$.

Table 3.4. *Axioms: Rules on Hierarchies* I. *The axioms apply to rules* $C : \mathcal{X}^k \longrightarrow \mathcal{X}$ *with* $\mathcal{X} \in \{\mathcal{H}, \mathcal{W}, \mathcal{W}_c\}$. *For notation see Definitions* 3.41 *on page* 42 *and* 3.49.

***Dct*: Dictatorship** $(\exists j \in K)(\forall X \subseteq S)(\forall P \in \mathcal{X}^k)(X \in H_j \Longrightarrow X \in CP)$

***Ind*: Independence** $(\forall X \subseteq S)(\forall P, P' \in \mathcal{X}^k)(P|_X = P'|_X \Longrightarrow CP|_X = CP'|_X)$

***PO*: Pareto Optimality** $(\forall X \subseteq S)(\forall P \in \mathcal{X}^k)(X \in H_K \Longrightarrow X \in CP)$

***Prj*: Projection** $(\exists j \in K)(\forall P \in \mathcal{X}^k)(CP = H_j)$

***RI*: Removal Independence** $(\forall X \subseteq S)(\forall P, P' \in \mathcal{X}^k)$
$(P|_X - X = P'|_X - X \Longrightarrow CP|_X - X = CP'|_X - X)$

***RTI*: Removal Ternary Independence** $(\forall X \subseteq S : |X| = 3)(\forall P, P' \in \mathcal{X}^k)$
$(P|_X - X = P'|_X - X \Longrightarrow CP|_X - X = CP'|_X - X)$

***Sym*: Symmetry**
$(\forall P \in \mathcal{X}^k)(\forall \text{ permutations } \sigma \text{ of } K)(CP = C(H_{\sigma(1)}, \ldots, H_{\sigma(k)}))$

***TPO*: Ternary Pareto Optimality**
$(\forall x, y, z \in S)(\forall P \in \mathcal{X}^k)(xyR_K z \Longrightarrow xyRz)$

***WI*: Weak Independence**
$(\forall X \subseteq S)(\forall P, P' \in \mathcal{X}^k)(P|_X = P'|_X \Longrightarrow CP|_X - X = CP'|_X - X)$

Consider what axioms would be suitable to describe consensus rules on hierarchies. In view of Arrow's success with weak orders, again it is natural to formulate axioms of independence, Pareto optimality, and dictatorship. In fact, the formulations in Table 3.4 of five of the axioms for hierarchies are simple restatements of corresponding axioms for equivalence relations in Table 3.1 on page 29. With them Barthélemy and McMorris [50] attempted to follow an Arrovian paradigm by stating for hierarchies that *Dct* is a consequence of *Ind* and *PO*; but in the presence of *PO*, *Ind* is too weak to imply *Dct*.

Example 3.50. [51, p. 44]. Let $C : \mathcal{H}^k \longrightarrow \mathcal{H}$ be a consensus rule. For all $P = (H_1, \ldots, H_k) \in \mathcal{H}^k$ let

$$CP = \begin{cases} H_\emptyset & \text{if } H_1 = H_\emptyset \text{ or } H_2 = H_\emptyset, \quad \text{otherwise} \\ \bigcup\{H_1|_Y : \text{ cluster } Y \text{ is maximal in } H_2\}. \end{cases}$$

One can verify that C is well defined and satisfies *Ind*. To see that *PO* holds, suppose $X \in H_K$; then a maximal cluster $Y \in H_2$ exists such that $X \subseteq Y$, so that $X = X \cap Y \in H_1|_Y$, whence $X \in CP$. Concerning *Dct*, suppose $S = \{1, \ldots, n\}$ for $n \geq 4$ and let $P \in \mathcal{H}^k$ be such that $H_1^* = \{\{1, \ldots, n-1\}\}$ and $H_i^* = \{\{2, \ldots, n\}\}$ for all $i \in K \setminus \{1\}$; then $CP^* = \{\{2, \ldots, n-1\}\}$, so no dictator exists. ■

How might *Ind* be strengthened to obtain an Arrow-like result? Removal independence (*RI*) implies *Ind* and so is stronger than *Ind*. To appreciate its relevance let $P, P' \in \mathcal{H}^k$ with $H = CP$ and $H' = CP'$, and suppose $P|_{xyz} = P'|_{xyz}$. Even though $xyRz$ and $xyR'z$, nevertheless it may be that $H|_{xyz} \neq H'|_{xyz}$: ternary independence, i.e., the restriction of *Ind* to ternary sets, need not be consistent with the equality of ternary representations. But when $xyRz$ and $xyR'z$ we have always that $H|_{xyz} - xyz = H'|_{xyz} - xyz$: removal ternary independence (*RTI*) is consistent with the equality of ternary representations, whereas ternary independence is not.

Example 3.51. For $S = wxyz$ let $H^* = \{xy, wz\}$ and $H'^* = \{xy, xyz\}$. Then $xyRz$, $xyR'z$, and $H|_{xyz} - xyz = H_\emptyset \cup \{xy\} = H'|_{xyz} - xyz$, whereas $H|_{xyz} = H_\emptyset \cup \{xy\} \neq H_\emptyset \cup \{xy, xyz\} = H'|_{xyz}$. ■

3.4.1 Barthélemy, McMorris, and Powers's Theorem

By using *RI* instead of *Ind*, Barthélemy, McMorris, and Powers obtain an impossibility theorem for strong hierarchies. Its import is that nondictatorships, which are desirable, are impossible in the presence of other equally desirable properties.

Theorem 3.52 (Barthélemy, McMorris, and Powers). [51, p. 45]. *Let C be a consensus rule on $\mathcal{H}$. C satisfies Dct if it satisfies RI and PO.*

We need several simple properties of ternary representations of hierarchies.

Proposition 3.53. [50, pp. 75–76]. *For all $H \in \mathcal{H}$ with $R = t(H)$:*

(i) $(\forall x, y, z \in S)(xyRz \iff xy \in H|_{xyz})$.
(ii) $(\forall Y \subset S)(Y \in H \setminus \{S\} \iff (\forall x, y \in Y)(\forall z \in S \setminus Y)(xyRz))$.
(iii) $(\forall x, y, \hat{y}, z \in S)(xyRz \wedge x\hat{y}Rz \implies y\hat{y}Rz)$.
(iv) $(\forall x, y, z, \hat{z} \in S)(xyRz \wedge z\hat{z}Rx \implies xyR\hat{z})$.

In Table 3.4, axioms *RTI* and *TPO* are ternary analogues of *RI* and *PO* and follow as consequences of them.

Lemma 3.54. [51, p. 44]. *Let C be a consensus rule on $\mathcal{H}$. If C satisfies RI and PO, then it satisfies RTI and TPO.*

Proof. Clearly $RI \implies RTI$. Let $P = (H_1, \dots, H_k) \in \mathcal{H}^k$ and $X = xyz \subseteq S$. In this proof only let $P^\star$, $H_i^\star$, etc., denote $P|_X - X$, $H_i|_X - X$, etc. Since $H_i^{\star\star} = H_i^\star$, then $P^{\star\star} = P^\star$, whence $C(P^\star)^\star = C(P)^\star$ by *RTI*. To establish *TPO* let xyR_Kz; then $xy \in H_K^\star$ by Proposition 3.53(i), so $xy \in C(P^\star)$ by *PO*, whence $xy \in C(P^\star) \implies xy \in C(P^\star)^\star \implies xy \in C(P)^\star$, and thus $xyRz$ by Proposition 3.53(i). □

Familiar concepts of decisiveness are relevant.

Definition 3.55. *Given a consensus rule $C : \mathcal{H}^k \longrightarrow \mathcal{H}$, let $I \subseteq K$ and let $xy|z$ be a triad with distinct $x, y, z \in S$. I is called* almost decisive for $xy|z$, *a condition denoted by $\widehat{U}^I_{xy|z}$, if every $P = (H_1, \ldots, H_k) \in \mathcal{H}^k$ having xyR_Iz and $xyzR_{K\setminus I}$ then implies $xyRz$. I is called* almost decisive *if it is almost decisive for all resolved triads. $\widehat{U}_C$ is the set of all almost decisive sets. I is called* decisive for $xy|z$, *a condition denoted by $U^I_{xy|z}$, if every $P = (H_1, \ldots, H_k) \in \mathcal{H}^k$ having xyR_Iz then implies $xyRz$. I is called* decisive *if it is decisive for all resolved triads. U_C is the set of all decisive sets.*

Lemma 3.56 establishes a requirement of invariant decisiveness.

Lemma 3.56. [51, p. 45]. *Let C be a consensus rule on $\mathcal{H}$ that satisfies RTI and TPO. If $I \subseteq K$, then*

$$(\exists abc \in S^3)(\widehat{U}^I_{ab|c}) \Longleftrightarrow I \in U_C.$$

Proof. The proof has five steps. (i) Assume $\widehat{U}^I_{xy|z}$ and let $t \in S$ be such that $t \notin \{x, y, z\}$; then we will show that $\widehat{U}^I_{xy|t}$. Since $|S| \geq 5$, construct $P \in \mathcal{H}^k$ to have $(\forall i \in I)(H_i^* = \{xy, tz\})$ and $(\forall i \in K \setminus I)(H_i^* = \{tz\})$. Since $\widehat{U}^I_{xy|z} \Longrightarrow xyRz$ and *TPO* $\Longrightarrow tzRx$, then $xyRt$ by Proposition 3.53(iv), whence *RTI* $\Longrightarrow \widehat{U}^I_{xy|t}$. (ii) Assume $\widehat{U}^I_{xy|z}$ and let $t \in S$ be such that $t \notin \{x, y, z\}$; then we will show that $\widehat{U}^I_{tx|z}$. Since $|S| \geq 5$, construct $P \in \mathcal{H}^k$ to have $(\forall i \in I)(H_i^* = \{txy\})$ and $(\forall i \in K \setminus I)(H_i^* = \{ty\})$. Since $\widehat{U}^I_{xy|z} \Longrightarrow xyRz$ and *TPO* $\Longrightarrow tyRz$, then $txRz$ by Proposition 3.53(iii), whence *RTI* $\Longrightarrow \widehat{U}^I_{tx|z}$. (iii) Using variations of (i) and (ii), we can obtain $\widehat{U}^I_{ab|c}$ for each triad $ab|c$ other than $xy|z$, whence $I \in \widehat{U}_C$. (iv) Let $P \in \mathcal{H}^k$ have xyR_Iz, where $X = \{x, y, z\} \subseteq S$; we will show that $xyRz$, whence $I \in U_C$. Since $|S| \geq 5$, let $t, w \notin X$ be distinct and construct $P' \in \mathcal{H}^k$ to have $(\forall i \in I)(H_i'^* = \{txy, wz\})$ and $(\forall i \in K \setminus I)(H_i' = H_i|_{xyz})$. Since $I \in \widehat{U}_C$ by step (iii), then $txR'w$, $tyR'w$, and $wzR't$, so that $xyR'z$ by Proposition 3.53(iii)–(iv). Since $P'|_X - X = P|_X - X$, then *RTI* $\Longrightarrow (CP'|_X - X = CP|_X - X)$, so $xyR'z \Longrightarrow xyRz$, whence $U^I_{xy|z}$ and thus $I \in U_C$. (v) The converse is trivial. □

With Lemma 3.56 we obtain this proof.

***Proof of Barthélemy, McMorris, and Powers's Theorem* 3.52.** Let rule $C : \mathcal{H}^k \longrightarrow \mathcal{H}$ satisfy *RI* and *PO*, so that *RTI* and *TPO* hold by Lemma 3.54. The main task is to prove that $(\exists j \in K)(\{j\} \in \widehat{U}_C)$, for then $(\exists j \in K)(\{j\} \in U_C)$ by Lemma 3.56, whence $(\forall X \subseteq S)(X \in H_j \Longrightarrow X \in CP)$ by Proposition 3.53(ii), and we are done. To begin, since *PO* $\Longrightarrow K \in \widehat{U}_C$, let $I \subseteq K$ be an almost decisive set of minimum cardinality. If $|I| > 1$, then a contradiction arises in the following way. Choose $j \in I$ arbitrarily, and for distinct $s, w, x, y, z \in S$, construct $P \in \mathcal{H}^k$ to have $H_j^* = \{xy, wz\}$, $(\forall i \in I \setminus \{j\})(H_i^* = \{sxy\})$ and $(\forall i \in K \setminus I)(H_i^* = \emptyset)$. Since xyR_Iz and $xyzR_{K\setminus I}$, then $I \in \widehat{U}_C \Longrightarrow xyRz$. Depending on the hierarchical structure of CP, we then must have either $syRz$ or $xyRs$; but the structure of P is such that $syRz \Longrightarrow \widehat{U}^{I\setminus\{j\}}_{sy|z}$ and $xyRs \Longrightarrow \widehat{U}^{\{j\}}_{xy|s}$, both of which are false. Thus assuming $|I| > 1$ is contradictory, so $(\exists j \in K)(\{j\} \in \widehat{U}_C)$ as was to be proved. □

Are the dictatorships arising from Theorem 3.52 on page 46 strong? Barthélemy, McMorris, and Powers answer this question for dictatorships that are weakly independent, where $RI \Longrightarrow Ind \Longrightarrow WI$.

Theorem 3.57. [52, p. 61]. *Let C be a consensus rule on $\mathcal{H}$. C satisfies Prj if it satisfies Dct and WI.*

With Theorem 3.52 on page 46 this yields characterizations of projections.

Theorem 3.58. [52, p. 63]. *Let C be a consensus rule on $\mathcal{H}$. C satisfies RI and PO if and only if C satisfies Prj.*

Proof. C is dictatorial by Theorem 3.52 on page 46. Since C satisfies *RI* and $RI \Longrightarrow WI$, then C is a projection by Theorem 3.57. The converse is trivial. □

Corollary 3.59. [52, p. 63]. *Let C be a consensus rule on $\mathcal{H}$ that satisfies Dct. These are equivalent:* (i) *C satisfies Prj.* (ii) *C satisfies WI.* (iii) *C satisfies Ind.* (iv) *C satisfies RI.*

Thus independence has not one but several natural analogues when translated from weak orders to hierarchies. Indeed Barthélemy, McMorris, and Powers [53] have investigated eight independence axioms for hierarchies including *Ind*, *RI*, and *WI*.

3.5 Weak Hierarchies

Perhaps less relevant to systematic biology is the case where the clusters of a hierarchy may overlap. Even so, 35 years ago Jardine and Sibson [210, 211, 212, 380] characterized the B_k clustering methods in which clusters may overlap by at most $k-1$ elements. Batbedat [58, 59] and Bandelt and Dress [28] also weaken the concept of strong hierarchy so as to allow overlapping of clusters (Figure 3.7).

Definition 3.60. *A* weak hierarchy *on S is a set $W \subseteq 2^S$ such that $S \in W$, $(\forall x \in S)(\{x\} \in W)$, and*

$$(\forall X, Y, Z \in W)(X \cap Y \cap Z \in \{X \cap Y, Y \cap Z, X \cap Z\}). \tag{3.3}$$

Each $X \in W$ is called a cluster. *$\mathcal{W}$ is the set of all weak hierarchies on S.*

Example 3.61. If $S = abcd$, then the rooted graph G_r in Figure 3.7 on the next page is representable by the weak hierarchy $W_1 = W_1^* \cup H_\emptyset$, where $W_1^* = \{ab, bc, cd, abc, bcd\}$ and clusters overlap, e.g., $|ab \cap bc| = 1$ and $|abc \cap bcd| = 2$. But any set $W_2 = W_2^* \cup H_\emptyset$ where $\{ab, ac, bc\} \subseteq W_2^*$ violates (3.3) and so is not a weak hierarchy. ■

Consider consensus rules $C : \mathcal{W}^k \longrightarrow \mathcal{W}$ that are subject to Convention 3.45 on page 44. For example, Bandelt and Dress [28] generalized the majority rule of Margush and McMorris [246] from strong hierarchies to mixtures of weak and strong hierarchies.

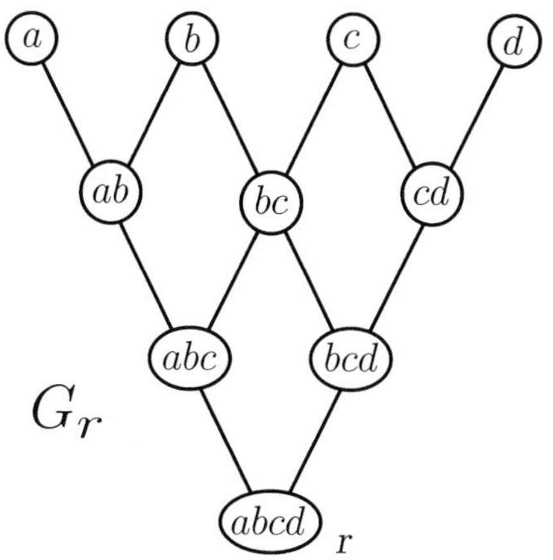

Figure 3.7. *Weak Hierarchy with Overlapping Clusters* [28, p. 146]

Definition 3.62. [28, p. 149] *The method of* weak majority rule *for weak hierarchies is the function* $Maj : \mathcal{W}^k \longrightarrow \mathcal{W}$ *such that, for* $P = (W_1, \ldots, W_l, W_{l+1}, \ldots, W_k) \in \mathcal{W}^k$ *with* $W_1, \ldots, W_l$ *being weak and* $W_{l+1}, \ldots, W_k$ *being strong,*

$$Maj\,P = \bigcup_{\substack{J \subseteq K \\ |J| > \frac{k+l}{3}}} \bigcap_{i \in J} W_i.$$

This rule is well defined.

Theorem 3.63. [28, p. 149]. *For all* $P = (W_1, \ldots, W_k) \in \mathcal{W}^k$ *with* $W_1, \ldots, W_l$ *being weak and* $W_{l+1}, \ldots, W_k$ *being strong,* $Maj\,P$ *is a weak hierarchy.*

Proof. If clusters $X_1, X_2, X_3 \in Maj\,P$ violate (3.3), then simple counting arguments yield a contradiction. By Definition 3.62, each $X_i \in Maj\,P$ results from the occurrences of more than $\frac{k+l}{3}$ clusters in the hierarchies of P, whence there are more than $k + l$ relevant occurrences for the three X_i; but the l weak hierarchies can account for at most $2l$ occurrences, and the $k - l$ strong hierarchies can account for at most $k - l$, whence there are at most $k + l$ relevant occurrences. □

Thus given any profile of k strong hierarchies, the clusters appearing in more than one-third of those hierarchies form a weak hierarchy; given any profile of k weak hierarchies, the clusters appearing in more than two-thirds of those weak hierarchies form a weak hierarchy. Moreover a characterization of the weak majority rule for weak hierarchies follows from Corollary 4.18 on page 58.

3.5.1 Powers's Theorem

An impossibility result exists for a type of weak hierarchy.

Definition 3.64. *A weak hierarchy $W \in \mathcal{W}$ on S is called* closed *if* $(\forall X, Y \in W)(X \cap Y \neq \emptyset \Longrightarrow X \cap Y \in W)$. *$\mathcal{W}_c$ is the set of all closed weak hierarchies on S, where $\mathcal{H} \subset \mathcal{W}_c \subset \mathcal{W}$.*

Consider consensus rules $C : \mathcal{W}_c^k \longrightarrow \mathcal{W}_c$ that are subject to this convention.

Convention 3.65. *To any (multi)consensus rule C with domain $\mathcal{W}_c^k$ is associated a set $S = S_n$ of n leaf labels on which the hierarchies are defined. In this context, unless specifically stated otherwise, S is finite with $|S| = n \geq 6$.*

By applying the axioms of Table 3.4 on page 45 to consensus rules on $\mathcal{W}_c$, Powers established for them an impossibility theorem that is based on weak independence. Since *RI* $\Longrightarrow$ *Ind* $\Longrightarrow$ *WI*, all three axioms yield impossibility results for rules on $\mathcal{W}_c$, a situation quite unlike that for rules on $\mathcal{H}$ (Theorem 3.52 on page 46), where only *RI* yields an impossibility result. Nondictatorships, which are desirable, are impossible for rules on $\mathcal{W}_c$ in the presence of other equally desirable properties.

Theorem 3.66 (Powers). [335, pp. 272]. *Let C be a consensus rule on $\mathcal{W}_c$. C satisfies Prj if it satisfies WI and PO.*

Powers's proof of this result can be summarized beginning with the following proposition.

Proposition 3.67. *If $W \in \mathcal{W}_c$ has the ternary representation $R = t(W)$, then*

(i) $(\forall x, y, z \in S)(xy \in W|_{xyz} \Longleftrightarrow xyRz)$.
(ii) $(\forall Y \subset S)(Y \in W \Longleftrightarrow (\forall x, y \in Y)(\forall z \in S \setminus Y)(xyRz))$.

As with strong hierarchies, *TPO* is the natural ternary analogue of *PO*.

Proposition 3.68. [335, lemma 1]. *If the rule $C : \mathcal{W}_c^k \longrightarrow \mathcal{W}_c$ satisfies WI and PO, then it satisfies TPO.*

Familiar concepts of decisiveness are relevant.

Definition 3.69. *Given a consensus rule $C : \mathcal{W}_c^k \longrightarrow \mathcal{W}_c$, let $I \subseteq K$ and let $xy|z$ be a triad with distinct $x, y, z \in S$. I is called* almost decisive *for $xy|z$, a condition denoted by $\widehat{U}^I_{xy|z}$, if every $P = (W_1, \ldots, W_k) \in \mathcal{W}_c^k$ having*

$$(\forall i \in I)(W_i|_{xyz} = H_\emptyset \cup \{xyz, xy\}) \text{ and}$$
$$(\forall i \in K \setminus I)(W_i|_{xyz} = H_\emptyset \cup \{xyz\})$$

then implies $xy|z \in CP$, i.e., $xyRz$. I is called almost decisive *if it is almost decisive for all resolved triads. $\widehat{U}_C$ is the set of all almost decisive sets.*

For each rule C satisfying *WI* and *PO*, Powers then proves

(i) [335, lemma 6]. $(\exists j \in K)(\{j\} \in \widehat{U_C})$, so that j is a dictatorial candidate.

(ii) [335, lemma 10]. $(\forall P \in \mathcal{W}_c^k)(\forall x, y, z \in S)(xyR_jz \implies xyRz)$, so that C satisfies *Dct*.

(iii) [335, lemma 14]. $(\forall P \in \mathcal{W}_c^k)(\forall x, y, z \in S)(xyRz \implies xyR_jz)$, so that C satisfies *Prj* and j is strongly dictatorial.

Since C satisfies *Ind* or *RI* only if C satisfies *WI*, we obtain the following corollary.

Corollary 3.70. [335, pp. 277]. *Let C be a consensus rule on $\mathcal{W}_c$ that satisfies PO. These are equivalent:* (i) *C satisfies Prj.* (ii) *C satisfies WI.* (iii) *C satisfies Ind.* (iv) *C satisfies RI.*

Open Problem 3.71. Determine if other reasonable concepts of independence or Pareto optimality give rise to impossibility theorems for consensus rules on $\mathcal{W}$ or $\mathcal{W}_c$. ■

3.6 Notes

In biology, Arrow's impossibility theorem on weak orders is relevant to the aggregation of evolutionary orders (McGuire and Thompson [252]) or of rankings of fitness (Cohen [134]). The *Journal of Classification*'s 1986 special issue on consensus classifications (Day [150]) demonstrates the interest at that time in biological consensus problems, e.g., [3, 48, 49, 183, 315, 392]. Swofford et al. [400] and Felsenstein [161] focus on phylogenetic inference, a branch of systematic biology where consensus hierarchies and phylogenies are used to estimate evolutionary relationships.

For section 3.1 (Partitions of a Set), see Mirkin [282, 283], Leclerc [233], Fishburn and Rubinstein [168], and also Barthélemy's [43] assessment of these contributions. Leclerc's general results on valued relations yield Mirkin's theorem as a special case [233, p. 54]. Rubinstein and Fishburn [353] describe an algebraic aggregation theory in which consensus problems for equivalence relations can be studied and Mirkin's theorem can be formulated. Barthélemy and Leclerc [47] investigate the median procedure for partitions of a set. The oligarchic concept (Definition 3.7 on page 29) applies of course to SWFs: Guha [189] and Mas-Colell and Sonnenschein [247] independently prove an oligarchic analogue of Arrow's theorem for SWFs in the quasi-transitive case where transitivity of weak orders is relaxed by imposing it for strict preference but not for indifference; Vincke [406] and Bouyssou [105] show that independent, Pareto optimal multiconsensus SWFs satisfy analogues of weak dictatorship and oligarchy.

As for the remaining sections, Leclerc [238] surveys the literature on the consensus of classification trees, including phylogenies and hierarchies. In the early literature, e.g., [246, 314], a hierarchy H on S_n was called an *n-tree*. For section 3.2 (Tree Quasi-orders), see McMorris and Neumann [267]. For section 3.3 (Phylogenies), see Colonius and Schulze [136], McMorris [259], and McMorris and Powers [269]. For section 3.4 (Hierarchies), see McGuire [251], Margush and McMorris [246], Barthélemy and McMorris [50], Barthélemy, McMorris, and Powers [51, 52, 53], and Powers [336]. Dwyer, McMorris, and Powers [157] investigate the case where for all $P = (H_1, \ldots, H_k) \in \mathcal{H}^k$ a multiconsensus rule C returns a set CP of one or two hierarchies: if C is removal independent and Pareto optimal, then some individual $j \in K$ dictates in the weak sense that $(\forall P \in \mathcal{H}^k)(H_j \subseteq \bigcup CP)$. In section 3.5 (Weak Hierarchies), the development is based on Bandelt and Dress [28] and Powers [335].

Biologists say too much, imprecisely; mathematicians say too little, but very precisely. Strive for the middle ground: say just enough with reasonable precision. — Anonymous

Chapter 4
Possibilities in Bioconsensus

Yet the single most important principle of cladistics is that diverse fundamental cladograms may be combined to form a single general cladogram.
— G. Nelson [313, p. 7]

I have myself been unable to find a better method than this even after much effort; and you can safely take it that a more perfect method cannot be found.
— N. Cusanus in *De Concordantia Catholica* (c. 1434) [257, p. 106]

In Chapter 3 we showed in bioconsensus how axiomatic methods establish impossibilities, i.e., impossibility theorems, which state that no consensus rule can exhibit various sets of desirable properties. Now we address in bioconsensus the more optimistic topic of how axiomatic methods establish possibilities, i.e., characterization theorems, which identify consensus rules having unique sets of desirable properties. Because they are relevant to systematics, taxonomy, and ecology, we will feature consensus rules for hierarchies (Definitions 3.40 on page 41 and 3.60 on page 48). Section 4.1 considers consensus rules having simple consensus criteria based on the frequencies with which clusters occur in a profile's hierarchies. Section 4.2 describes more complex rules that may put a cluster in the consensus hierarchy even if it is in none of the profile's hierarchies. Section 4.3 concerns the median multiconsensus rules, which minimize a distance-based measure of remoteness and so must accommodate the existence of one or more consensus hierarchies.

4.1 Counting Rules

By now the reader may think that, for typical sets $\mathcal{X}$ of structures on S, the axioms of independence and Pareto optimality are too strong to admit other than trivial or undesirable consensus rules. But if *Ind* or *PO* must be abandoned, what axioms might comprise a weaker axiom set that would characterize one or more useful rules? Let each element of each structure in $\mathcal{X}$ be a nonempty subset of S that is called a cluster. Consider rules

Table 4.1. *Axioms: Rules on Hierarchies* II. *The axioms apply to rules* $C : \mathcal{X}^k \longrightarrow \mathcal{X}$ *with* $\mathcal{X} \in \{\mathcal{H}, \mathcal{W}, \mathcal{W}_c\}$. *For notation see Definition* 3.49 *on page* 44.

***Atn*: Autonomy** $(\forall \emptyset \neq X \subset S)(\exists P \in \mathcal{X}^k)(X \in CP)$

***CPO*: co-Pareto Optimality** $(\forall P \in \mathcal{X}^k)(CP \subseteq \bigcup_{i=1}^k H_i)$

***Dct*: Dictatorship** $(\exists j \in K)(\forall X \subseteq S)(\forall P \in \mathcal{X}^k)(X \in H_j \Longrightarrow X \in CP)$

***DM*: Decisive Monotonicity**

$$(\forall X \subseteq S)(\forall P, P' \in \mathcal{X}^k)(K_X(P) \subseteq K_X(P') \Longrightarrow [X \in CP \Longrightarrow X \in CP'])$$

***DN*: Decisive Neutrality**

$$(\forall X, Y \subseteq S)(\forall P, P' \in \mathcal{X}^k)(K_X(P) = K_Y(P') \Longrightarrow [X \in CP \Longleftrightarrow Y \in CP'])$$

***Ind*: Independence** $(\forall X \subseteq S)(\forall P, P' \in \mathcal{X}^k)(P|_X = P'|_X \Longrightarrow CP|_X = CP'|_X)$

***MN*: Monotonic Neutrality**

$$(\forall X, Y \subseteq S)(\forall P, P' \in \mathcal{X}^k)(K_X(P) \subseteq K_Y(P') \Longrightarrow [X \in CP \Longrightarrow Y \in CP'])$$

***PO*: Pareto Optimality** $(\forall X \subseteq S)(\forall P \in \mathcal{X}^k)(X \in H_K \Longrightarrow X \in CP)$

***Sym*: Symmetry** $(\forall P \in \mathcal{X}^k)(\forall \text{ permutations } \sigma \text{ of } K)(CP = C(H_{\sigma(1)}, \ldots, H_{\sigma(k)}))$

$C : \mathcal{X}^k \longrightarrow \mathcal{X}$ in which, for all $P \in \mathcal{X}^k$, a cluster is put in $C(P) = CP$ if it appears sufficiently often in P's structures.

4.1.1 Strong Hierarchies

To focus the discussion further, let $\mathcal{H}$ be the set of all strong hierarchies on S, consider consensus rules $C : \mathcal{H}^k \longrightarrow \mathcal{H}$ and view a hierarchy's nontrivial clusters as its defining entities. Since the impossibility results of Chapter 3 depend on decisive sets and such related properties as invariant decisiveness and equivalent subsets, one might ask: What properties of decisiveness are desirable? If substituted for *Ind* and *PO*, what axioms would retain a viable concept of decisive set? Two axioms spring to mind: a form of neutrality strong enough to yield invariant decisiveness and a form of monotonicity strong enough to ensure that every superset of a decisive set is decisive. Table 4.1 lists candidates. For decisive neutrality (*DN*) suppose cluster X is in P at exactly the same positions as Y is in P'; then X should be in CP if and only if Y is in CP'. For decisive monotonicity (*DM*), if cluster X is in CP and if we change P to P' by putting X in more profile hierarchies, then X should also be in CP'. Autonomy (*Atn*) ensures that the individuals in society are free to choose among all clusters: for each cluster X there should be a profile $P \in \mathcal{H}^k$ such that X is in CP. Symmetry (*Sym*) ensures the anonymity or equality of individuals in society. Co-Pareto optimality (*CPO*) ensures that clusters don't come out of nowhere: each cluster in CP should be in at least one of P's hierarchies. Several relationships among these axioms are easy to prove.

Proposition 4.1. *Let C be a consensus rule on $\mathcal{H}$.*

(i) *C satisfies DM and DN if and only if C satisfies MN.*
(ii) *C satisfies CPO if C satisfies MN.*
(iii) *C satisfies PO if C satisfies MN and Atn.*

Just as some consensus rules are related to filters or ultrafilters (Definition 2.21 on page 22), so others are related to variants of the filter concept.

Definition 4.2. [267, p. 135] *A set $\mathcal{D} \subseteq 2^K$ is a* semidecisive family *on K if*

$$(\forall I, J \in \mathcal{D})(I \cap J \neq \emptyset). \tag{4.1a}$$

A set $\mathcal{D} \subseteq 2^K$ is a decisive family *on K if it is semidecisive and*

$$(\forall I \in \mathcal{D})(\forall J \subseteq K)(I \subseteq J \Longrightarrow J \in \mathcal{D}). \tag{4.1b}$$

Condition (4.1a) relaxes the requirement (2.2c) that a filter be closed under set intersection, while (4.1b) is the monotonicity requirement (2.2b) of filters. Although every filter is a decisive family, the converse is false: if $K = 123$, then $\mathcal{D} = \{12, 13, 23, 123\}$ is a decisive family on K, yet $\mathcal{D}$ is not a filter since intersection invariance is violated. Although every decisive family is semidecisive, the converse is false: if $K = 12345$, then $\mathcal{D} = \{123, 124, 125, 134, 135, 145, 234, 235, 245, 345\}$ is a semidecisive family on K, yet $\mathcal{D}$ is not decisive since monotonicity is violated.

Example 4.3. The following are decisive families:

(i) If $i \in K$, then $\mathcal{D}_{\{i\}} = \{J \subseteq K : i \in J\}$ is called a *dictatorial family* on K. The ultrafilters on a finite set K are the dictatorial families on that set.
(ii) If $\emptyset \neq I \subseteq K$, then $\mathcal{D}_I = \{J \subseteq K : I \subseteq J\}$ is called an *oligarchic family* on K. The filters on a finite set K are the oligarchic families on that set.
(iii) If $\ell \in K$ with $\ell > \frac{k}{2}$, then $\mathcal{D}_\ell = \{J \subseteq K : \ell \leq |J|\}$ is a *quota family* on K. ■

Familiar concepts of decisiveness can be related to semidecisive families.

Definition 4.4. *Given a consensus rule $C : \mathcal{H}^k \longrightarrow \mathcal{H}$, let $I \subseteq K$ and let $X \subseteq S$ be a nonempty cluster. I is called* decisive for X, *a condition denoted by U_X^I, if every $P \in \mathcal{H}^k$ having $X \in H_I$ then implies $X \in CP$. I is called* decisive *if it is decisive for all nonempty clusters. U_C is the set of all decisive sets.*

Lemma 4.5 establishes a familiar requirement of invariant decisiveness.

Lemma 4.5. *Let C be a consensus rule on $\mathcal{H}$ that satisfies DN. If $I \subseteq K$, then*

$$(\exists X \subseteq S)(U_X^I) \Longleftrightarrow I \in U_C.$$

Proof. Let $I \subseteq K$ and U_X^I for $X \subseteq S$, so that $P \in \mathcal{H}^K$ exists with $X \in H_I$ and $X \in CP$. Consider any $Y \subseteq S$ and $P' \in \mathcal{H}^K$ such that $Y \in H_I'$. Then $Y \in CP'$ by *DN*, whence U_Y^I, so $I \in U_C$. The converse is trivial. □

Consensus rules can be specified by semidecisive families.

Definition 4.6. *For each semidecisive family $\mathcal{S} \subseteq 2^K$ let $M_{\mathcal{S}} : \mathcal{H}^k \longrightarrow \mathcal{H}$ be the consensus rule such that*

$$(\forall P = (H_1, \ldots, H_k) \in \mathcal{H}^k)\left(M_{\mathcal{S}}P = \bigcup_{V \in \mathcal{S}} \bigcap_{i \in V} H_i\right).$$

Such rules are almost always well defined.

Lemma 4.7. *If $\emptyset \neq \mathcal{S} \subseteq 2^K$ is a semidecisive family and $P = (H_1, \ldots, H_k) \in \mathcal{H}^k$, then $M_{\mathcal{S}}P \in \mathcal{H}$.*

Proof. Since $\mathcal{S} \neq \emptyset$, then $S \in M_{\mathcal{S}}P$ and $(\forall x \in S)(\{x\} \in M_{\mathcal{S}}P)$. If $X, Y \in M_{\mathcal{S}}P$, then $I, J \in \mathcal{S}$ exist where $X \in \bigcap_{i \in I} H_i$ and $Y \in \bigcap_{i \in J} H_i$; but since $I \cap J \neq \emptyset$ by (4.1a), $\{X, Y\} \subseteq H_j$ for all $j \in I \cap J$, whence $X \cap Y \in \{\emptyset, X, Y\}$. □

Consensus rules based on semidecisive families can be characterized.

Theorem 4.8. Neumann [314, p. 286]. *Let C be a consensus rule on $\mathcal{H}$. C satisfies DN if and only if $C = M_{\mathcal{S}}$ for a semidecisive family $\mathcal{S}$.*

Proof. Let $C = M_{\mathcal{S}}$ for a semidecisive family $\mathcal{S}$. Let $P, P' \in \mathcal{H}^k$ and $X, Y \subseteq S$ such that $K_X(P) = K_Y(P')$. If $X \in M_{\mathcal{S}}P$, then $(\exists V \in \mathcal{S})\,(X \in \bigcap_{i \in V} H_i)$, where $V \subseteq K_X(P) = K_Y(P')$, so $Y \in \bigcap_{i \in V} H_i'$ and thus $Y \in M_{\mathcal{S}}P'$. Similarly $Y \in M_{\mathcal{S}}P' \implies X \in M_{\mathcal{S}}P$, whence $M_{\mathcal{S}}$ satisfies *DN*.

Let C satisfy *DN*. First we show that U_C is a semidecisive family. Since *DN* holds, U_C satisfies invariant decisiveness by Lemma 4.5. Since $|S| \geq 5$, suppose $a, b, c, d, e \in S$. Let $I, J \in U_C$; if $I \cap J = \emptyset$, then build $P = (H_1, \ldots, H_k) \in \mathcal{H}^k$ with $(\forall i \in I)(H_i^* = \{ab\})$, $(\forall i \in J)(H_i^* = \{bc\})$, and $(\forall i \in K \setminus (I \cup J))(H_i^* = \emptyset)$. But *DN* and invariant decisiveness ensure that $ab, bc \in CP$, violating for $CP \in \mathcal{H}$ the requirement that $ab \cap bc \in \{\emptyset, ab, bc\}$. Thus $I \cap J \neq \emptyset$, whence U_C satisfies (4.1a) and thus is a semidecisive family. Finally we show that $C = M_{U_C}$. If $X \in CP$ for some $P \in \mathcal{H}^k$, then $I = K_X(P) \in U_C$ by *DN*, so $(\exists I \in U_C)\,(X \in \bigcap_{i \in I} H_i \subseteq M_{U_C}P)$, whence $CP \subseteq M_{U_C}P$. If $X \in M_{U_C}P$, then $(\exists I \in U_C)\,(X \in \bigcap_{i \in I} H_i)$, so the decisiveness of I ensures that $X \in CP$, whence $M_{U_C}P \subseteq CP$. Thus $C = M_{U_C}$ for the semidecisive family U_C. □

Rules based on decisive families, i.e., *counting rules*, can also be characterized.

Theorem 4.9. McMorris and Neumann [267, p. 135]. *Let C be a consensus rule on $\mathcal{H}$. C satisfies MN and Atn if and only if $C = M_{\mathcal{D}}$ for a decisive family $\mathcal{D}$.*

Proof. Clearly $C = M_{\mathcal{D}}$ satisfies *MN* and *Atn* for each decisive family $\mathcal{D}$. For the converse, let C satisfy *MN* and *Atn*. By Theorem 4.8, $C = M_{\mathcal{U}_C}$ for the semidecisive family $\mathcal{U}_C$; we have only to show that $\mathcal{U}_C$ satisfies condition (4.1b) and so is a decisive family. Since $\mathcal{U}_C \neq \emptyset$ by *PO*, let $I \in \mathcal{U}_C$ with $I \subseteq J \subseteq K$; then *MN* and invariant decisiveness ensure that $J \in \mathcal{U}_C$, whence $\mathcal{U}_C$ satisfies (4.1b). □

By Example 4.3 on page 55 this result applies to dictatorships and oligarchies, and by imposing symmetry we obtain a characterization of the $M_{\mathcal{D}_\ell}$ family of *quota rules.*

Corollary 4.10. [267, p. 136]. *Let C be a consensus rule on* $\mathcal{H}$. *C satisfies MN, Atn, and Sym if and only if* $C = M_{\mathcal{D}_\ell}$ *for* $\ell > \frac{k}{2}$.

Proof. Let C satisfy *MN*, *Atn*, and *Sym*; then, by Theorem 4.9, $C = M_{\mathcal{D}}$ for a decisive family $\mathcal{D}$. By *Sym* the image of any decisive set under a permutation of K must also be decisive, so $\mathcal{D}$ consists of subsets with cardinality at least some fixed $\ell \in K$. Since decisive sets intersect nontrivially, then $\ell > \frac{k}{2}$ and $C = M_{\mathcal{D}_\ell}$. The converse is straightforward. □

$M_{\mathcal{D}_\ell}$ is the strict consensus rule when $\ell = k$; it is the majority rule when $\ell = \lceil \frac{k+1}{2} \rceil$, i.e., when ℓ is the smallest integer greater than $\frac{k}{2}$.

4.1.2 Weak Hierarchies

A challenging problem arises in bioconsensus or cluster analysis if proximity data are analyzed using both strong (Definition 3.40 on page 41) and weak (Definition 3.60 on page 48) hierarchies.

Open Problem 4.11. Let similarity or dissimilarity data be given for a study collection S and let a set K of clustering programs be used to analyze these data. For $L \subseteq K$ let the $l = |L|$ algorithms indexed by L yield weak hierarchies as output, while those indexed by $L^c = K \setminus L$ yield strong hierarchies. Describe rigorously whatever is in common agreement among these k hierarchies. ■

Of the many ways to attack this problem, the one used by McMorris and Powers [268] is to obtain a characterization (Theorem 4.17) by extending the results of section 4.1.1 from strong to weak hierarchies. Let $\mathcal{H} *_L \mathcal{W}$ be the set of all profiles $P = (H_1, \ldots, H_k) \in \mathcal{W}^k$ such that $(\forall i \in L^c)\,(H_i \in \mathcal{H})$; such a profile is called an L*-profile*. To accommodate L-profiles, symmetry (*Sym* in Table 4.1 on page 54) must be qualified so that for an L-profile P only those permutations σ are feasible which map P onto an L-profile P_σ. Notice that $\mathcal{H} *_\emptyset \mathcal{W} = \mathcal{H}^k$ and $\mathcal{H} *_K \mathcal{W} = \mathcal{W}^k$. For all $I \subseteq K$ one may distinguish the strong and weak hierarchies by setting $I_s = I \cap L^c$ and $I_w = I \cap L$. With this notation the concept of decisive family (Definition 4.2 on page 55) can be extended to mixtures of strong and weak hierarchies.

Definition 4.12. [268, p. 681]. *Any $\mathcal{D} \subseteq 2^K$ is an* L-weak decisive family *on K if*

$$(\forall I^1, I^2, I^3 \in \mathcal{D})(I^1_w \cap I^2_w \cap I^3_w \neq \emptyset \vee (\exists i \neq j \in \{1,2,3\})(I^i_s \cap I^j_s \neq \emptyset)) \text{ and}$$
$$(\forall I \in \mathcal{D})(\forall J \subseteq K)(I \subseteq J \Longrightarrow J \in \mathcal{D}).$$

Familiar concepts of decisiveness can be related to L-weak decisive families.

Definition 4.13. *Given for $L \subseteq K$ a consensus rule $C : \mathcal{H} *_L \mathcal{W} \longrightarrow \mathcal{W}$ on L-profiles, let $I \subseteq K$ and let $X \subseteq S$ be a nonempty cluster. I is called* decisive for X, *a condition denoted by U^I_X, if every $P \in \mathcal{H} *_L \mathcal{W}$ such that $X \in H_I$, then implies $X \in CP$. I is called* decisive *if it is decisive for all nonempty clusters. U_C is the set of all decisive sets.*

Lemma 4.14 establishes a familiar requirement of invariant decisiveness.

Lemma 4.14. *For $L \subseteq K$ let $C : \mathcal{H} *_L \mathcal{W} \longrightarrow \mathcal{W}$ be a consensus rule that satisfies DN. If $I \subseteq K$, then*

$$(\exists X \subseteq S)(U^I_X) \Longleftrightarrow I \in U_C.$$

Consensus rules can be specified by L-weak decisive families.

Definition 4.15. *For each L-weak decisive family $\mathcal{D} \subseteq 2^K$ let $M_\mathcal{D} : \mathcal{H} *_L \mathcal{W} \longrightarrow \mathcal{W}$ be the consensus rule such that*

$$(\forall P = (H_1, \ldots, H_k) \in \mathcal{H} *_L \mathcal{W})\left(M_\mathcal{D} P = \bigcup_{V \in \mathcal{D}} \bigcap_{i \in V} H_i\right).$$

Such rules are almost always well defined.

Lemma 4.16. [268, p. 681]. *If $\emptyset \neq \mathcal{D} \subseteq 2^K$ is an L-weak decisive family and $P = (H_1, \ldots, H_k) \in \mathcal{H} *_L \mathcal{W}$, then $M_\mathcal{D} P \in \mathcal{W}$.*

Analogues of Theorem 4.9 and Corollary 4.10 yield these characterizations of consensus rules based on L-weak decisive families.

Theorem 4.17. McMorris and Powers [268, p. 682]. *For $L \subseteq K$ let $C : \mathcal{H} *_L \mathcal{W} \longrightarrow \mathcal{W}$ be a consensus rule. C satisfies MN and Atn if and only if $C = M_\mathcal{D}$ for an L-weak decisive family $\mathcal{D}$.*

Corollary 4.18. [268, p. 682]. *For $L \subseteq K$ let $C : \mathcal{H} *_L \mathcal{W} \longrightarrow \mathcal{W}$ be a consensus rule. C satisfies MN, Atn, and Sym if and only if $C = M_{\mathcal{D}_\ell}$ for $\ell > \frac{k+l}{3}$.*

When $\ell = \lceil \frac{k+l+1}{3} \rceil$, Corollary 4.18 characterizes the weak majority rule for weak hierarchies (Definition 3.62 on page 49). And notice how the majority rule changes in

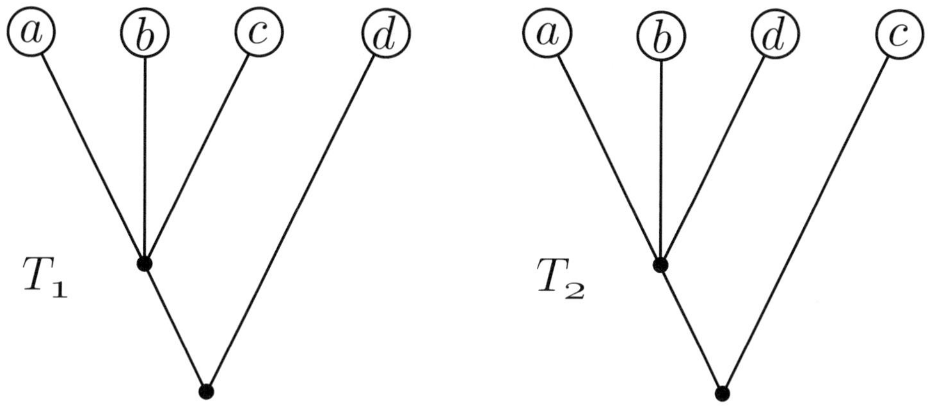

Figure 4.1. *Structure in Rooted Trees* I [3, p. 303]. *Example* 4.19 *explains.*

its transformation from strong to weak hierarchies [268, p. 683]. The original majority rule $Maj : \mathcal{H}^k \longrightarrow \mathcal{H}$ (Definition 3.47 on page 44) is a $\frac{1}{2}$-*rule* putting a cluster in the result if it appears in more than one-half of the profile's elements. Consider the majority rule $Maj : \mathcal{H}^k \longrightarrow \mathcal{W}$ in which the consensus hierarchy may be weak; set $l = 0$ in Corollary 4.18 to see that this Maj is a $\frac{1}{3}$-*rule* putting a cluster in the result if it appears in more than one-third of the profile's elements. Consider the majority rule $Maj : \mathcal{W}^k \longrightarrow \mathcal{W}$ in which all of the profile's hierarchies are weak; set $l = k$ in Corollary 4.18 to see that this Maj is a $\frac{2}{3}$-*rule* putting a cluster in the result if it appears in more than two-thirds of the profile's elements.

4.2 Intersection Rules

What structure is possessed and shared by rooted trees?

Example 4.19. Let T_1 and T_2 be rooted trees with labeled leaves (Figure 4.1). If we represent each T_i by a hierarchy H_i on $S = abcd$, then $H_1^* = \{abc\}$ and $H_2^* = \{abd\}$: since $H_1^* \cap H_2^* = \emptyset$, the trees appear to share no nontrivial structure. If we represent each T_i by its set R_i of triads, then $R_1 = \{abc, ab|d, ac|d, bc|d\}$ and $R_2 = \{abd, ab|c, ad|c, bd|c\}$: since $R_1 \cap R_2 = \emptyset$, the trees appear to share no nontrivial structure. Yet we have missed at least one shared structural feature: the leaf set ab joins in each tree at a greater height (farther from the root) than does S, which joins at the root. This *nesting of ab in S* (Definition 4.21 on page 61) is denoted by $ab < S$. If we represent each T_i by the set N_i of its nontrivial nestings, then $N_1 = \{ab < S, ac < S, bc < S, abc < S\}$ and $N_2 = \{ab < S, ad < S, bd < S, abd < S\}$: since $N_1 \cap N_2 = \{ab < S\}$, T_1 and T_2 appear to share this structural fragment. ■

In the example, $ab = abc \cap abd$ for $abc \in H_1^*$ and $abd \in H_2^*$: it is the set intersection of a cluster from each profile hierarchy. Were ab to be in CP for $P = (H_1, H_2)$, a cluster

ALGORITHM 4.1.
Adams Consensus Rule on Hierarchies [334, p. 54]
Input: $P' = (H_1', \ldots, H_k') \in \mathcal{H}^k$.
Output: $C_a^* P'$.
See (4.2a)–(4.2c) for specifications of $\max C_a P$, $V_i(P)$, and $P \setminus V(P)$.

begin
  $t \leftarrow \emptyset$
  $P \leftarrow P'$
  while ($\max C_a P \neq \emptyset$) **do**
    begin
      $t \leftarrow t \cup \max C_a P$
      $P \leftarrow P \setminus V(P)$
    end
  $C_a^* P' \leftarrow t$
end.

would be in a consensus hierarchy without it being in any profile hierarchy. Some consensus rules have this property. A consensus rule is called an *intersection* rule if it calculates the clusters of the consensus hierarchy by taking set intersections of k clusters, one from each profile hierarchy. Intersection rules differ fundamentally from the counting or quota rules of section 4.1.1: in the first instance a cluster of the consensus hierarchy may occur in none, one, or all of the profile hierarchies; in the second, it must occur in profile hierarchies with a frequency determined by the decisive family associated with the rule.

4.2.1 Adams's Rule

For $\mathcal{H}$ the set of all hierarchies on S, consider consensus rules $C : \mathcal{H}^k \longrightarrow \mathcal{H}$. If $H \in \mathcal{H}$, let $\max H$ be the set of all nontrivial maximal clusters of H with respect to set inclusion, and let H^* be the set of all nontrivial clusters that belong to H. Generally, if Z is a set of subsets of S, then $Z^* = \{B \in Z : 1 < |B| < |S|\}$. In 1972 E. N. Adams [2] proposed what became a widely used consensus rule for hierarchies. Algorithm 4.1 calculates the Adams consensus result $C_a P$ for each profile P. It identifies iteratively all the maximal nontrivial clusters that can be used to describe nestings shared by every hierarchy H_i in P. It works through the profile hierarchies in the direction away from their roots.

1. Calculate the maximal clusters shared by clusters in every H_i:
$$\max C_a P = \{\cap_{i=1}^k X_i : X_i \in \max H_i\}^*. \tag{4.2a}$$
2. If $\max C_a P \neq \emptyset$, then these clusters become part of the consensus result. In each H_i identify the nontrivial clusters $B \in H_i^*$ that contain any of the clusters in $\max C_a P$:
$$(\forall i \in K)(V_i(P) = \{B \in H_i^* : (\exists X \in \max C_a P)(X \subseteq B)\}). \tag{4.2b}$$
Since such clusters have yielded their information on shared nestings, delete them from the profile's hierarchies:
$$P \setminus V(P) = (H_1 \setminus V_1(P), \ldots, H_k \setminus V_k(P)). \tag{4.2c}$$
3. Repeat steps 1 and 2 until $\max C_a P = \emptyset$, when the algorithm terminates.

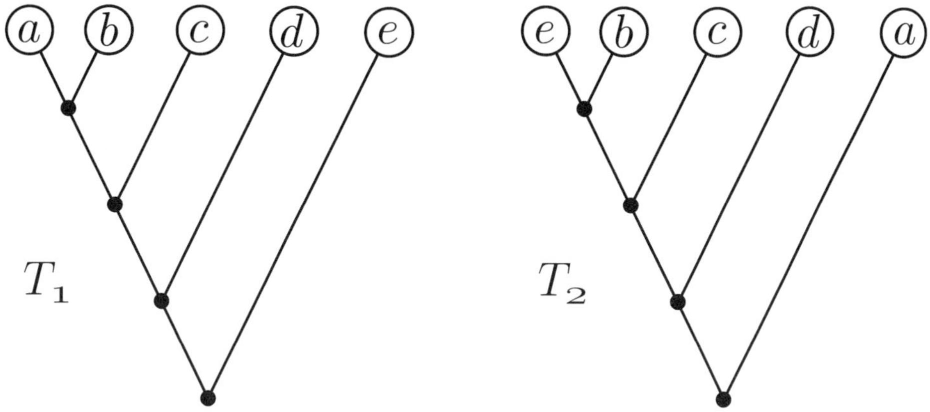

Figure 4.2. *Structure in Rooted Trees* II [314, p. 274]. *Example* 4.20 *explains.*

Example 4.20. [314, p. 278]. Let $P = (H_1, H_2)$ for $S = abcde$, $H_1^* = \{ab, abc, abcd\}$, and $H_2^* = \{be, bce, bcde\}$ (Figure 4.2). Algorithm 4.1 on the facing page executes the body of the while loop twice before the test fails. The values assigned to relevant variables are

Iteration	t	H_1^*	H_2^*	$\max C_a P$
–	$\emptyset$	$\{ab, abc, abcd\}$	$\{be, bce, bcde\}$	$\{bcd\}$
1	$\{bcd\}$	$\{ab, abc\}$	$\{be, bce\}$	$\{bc\}$
2	$\{bc, bcd\}$	$\{ab\}$	$\{be\}$	$\emptyset$

whence $C_a^* P = \{bc, bcd\}$ at termination. ■

Adams's original description of C_a raised questions about its definition [246, 263], questions which Adams answered by characterizing C_a (see Theorem 4.23 on page 63) in terms of nestings.

Definition 4.21. [3, p. 305]. *Let* $<_H$ *be a binary relation on* 2^S *such that* $(\forall i \in S)$ *and* $(\forall \emptyset \neq X, Y, Z \subseteq S)$

$$X <_H Y \implies X \subset Y, \tag{4.3a}$$

$$X \neq \{i\} \implies \{i\} <_H X \cup \{i\}, \tag{4.3b}$$

$$X <_H Z \wedge Y <_H Z \implies (X \cup Y <_H Z) \vee (X \cap Y = \emptyset), \tag{4.3c}$$

$$X <_H Y \implies (Z \subset X \implies Z <_H Y) \wedge (Y \subset Z \implies X <_H Z) \wedge (X \subset Z \subset Y \implies X <_H Z \vee Z <_H Y), \tag{4.3d}$$

where (4.3d) *can be replaced (Domenach and Leclerc* [156, §3.2]*) by the more compact*

$$X \subset Y \subset Z \implies [X <_H Z \iff (X <_H Y) \vee (Y <_H Z)]. \tag{4.3e}$$

Any ordered pair (X, Y) *for which* $X <_H Y$ *is called a* nesting, *where* X nests *in* Y.

Table 4.2. *Axioms: Rules on Hierarchies* III. *For notation see Definitions* 3.49 (*page* 44), 4.39 (*page* 69), *and* 4.42 (*page* 70).

***Btw*: Betweenness**
$(\forall P \in \mathcal{H}^k)(\forall i \in K \text{ with } X_i \in H_i)(\exists B \in CP)(\bigcap_{i=1}^k X_i \subseteq B \subseteq \bigcup_{i=1}^k X_i)$

***Btw$_a$*: a-Betweenness** $(\forall P \in \mathcal{H}^k)(CP \setminus \max CP \subseteq C(P \setminus V(P)) \subseteq CP)$

***Btw$_h$*: h-Betweenness** $(\forall P \in \mathcal{H}^k$ with height assignment h on $\mathcal{H}^k)$
$(CP \setminus \max CP \subseteq C(P \setminus U(P, h)) \subseteq CP)$

***DN*: Decisive Neutrality**
$(\forall X, Y \subseteq S)(\forall P, P' \in \mathcal{H}^k)(K_X(P) = K_Y(P') \Longrightarrow [X \in CP \Longleftrightarrow Y \in CP'])$

***Ind*: Independence** $(\forall X \subseteq S)(\forall P, P' \in \mathcal{H}^k)(P|_X = P'|_X \Longrightarrow CP|_X = CP'|_X)$

***NP*: Nesting Preservation**
$(\forall P \in \mathcal{H}^k)(\forall \emptyset \neq X, Y \subseteq S)((\forall i \in K)(X <_{H_i} Y) \Longrightarrow X <_{CP} Y)$

***PO*: Pareto Optimality** $(\forall X \subseteq S)(\forall P \in \mathcal{H}^k)(X \in H_K \Longrightarrow X \in CP)$

***Prj*: Projection** $(\exists j \in K)(\forall P \in \mathcal{H}^k)(CP = H_j)$

***QSP*: Qualified Strong Presence**
$(\forall P \in \mathcal{H}^k)(\forall X, Y \in CP)(X <_{CP} Y \Longrightarrow (\forall i \in K)(X <_{H_i} Y))$

***SP*: Strong Presence**
$(\forall P \in \mathcal{H}^k)(\forall \emptyset \neq X, Y \subseteq S)(X <_{CP} Y \Longrightarrow (\forall i \in K)(X <_{H_i} Y))$

***USP*: Upper Strong Presence**
$(\forall P \in \mathcal{H}^k)(\forall X \in CP)(X <_{CP} S \Longrightarrow (\forall i \in K)(X <_{H_i} S))$

Nesting relations are important because the set $\mathcal{H}$ of hierarchies on S is in one-to-one correspondence with the set of nesting relations $<_H$ on S by the rule

$$(\forall H \in \mathcal{H})(\forall \emptyset \neq X, Y \subseteq S)(X <_H Y \Longleftrightarrow X \subset Y \wedge lub_H(X) \subset lub_H(Y)),$$

$lub_H(X)$ being the smallest $Z \in H$ with $X \subseteq Z$ [3, p. 306] or, equivalently,

$$(\forall H \in \mathcal{H})(\forall \emptyset \neq X, Y \subseteq S)(X <_H Y \Longleftrightarrow X \subset Y \wedge (\exists Z \in H)(X \subseteq Z \wedge Y \nsubseteq Z)).$$

Consider what axioms of nestings would be suitable to describe consensus rules on hierarchies. Nesting preservation (*NP* in Table 4.2) is an analogue of Pareto optimality: if a nesting is in every hierarchy of profile P, then it should be in the consensus hierarchy for P. Strong presence (*SP*), which is the converse of *NP*, is desirable but not attainable in the presence of *NP*.

Example 4.22. [3, p. 310]. Consider the profile $P = (H_1, H_2) \in \mathcal{H}^2$ on $S = abcd$ such that $H_1^* = \{ab, abc\}$ and $H_2^* = \{bc, bcd\}$. Since $bc <_{H_i} bcd$ and $bc <_{H_i} S$ are the only nontrivial shared nestings, it would follow by *NP* that $bc \in C^*P$. But placing a in the consensus hierarchy then creates $bc <_{CP} abc$ or $abc <_{CP} S$; since neither nesting is shared by H_1 and H_2, *SP* is violated. ■

Qualified strong presence (*QSP*), which is a weak form of *SP*, requires only that the clusters of the consensus hierarchy engender nestings common to all hierarchies of the profile. *NP* and *QSP* suffice to characterize C_a.

Theorem 4.23 (Adams). [3, p. 312]. *Let C be a consensus rule on* $\mathcal{H}$. *C satisfies NP and QSP if and only if* $C = C_a$.

Readers may consult [3, p. 312–314] for Adams's proof of this result, which establishes for all $P \in \mathcal{H}^k$ that at most one consensus hierarchy in the standard form $H^* \cup H_\emptyset$ satisfies *NP* and *QSP*, and that the hierarchy obtained by Algorithm 4.1 on page 60 satisfies *NP* and *QSP*. Powers [334] obtains a different characterization (Theorem 4.43 on page 70) by viewing C_a as an instance of a faithful consensus rule.

4.2.2 Faithful Rules

Believing that consensus hierarchies should retain as much information as possible from a profile's hierarchies, Neumann [314] investigates consensus rules based on a concept of how a cluster best represents a given set of clusters.

Definition 4.24. *A consensus rule* $C : \mathcal{H}^k \longrightarrow \mathcal{H}$ *satisfies* betweenness (*Btw*) *and is called* faithful *if, for each* $P = (H_1, \ldots, H_k) \in \mathcal{H}^k$ *and each* $i \in K$ *with associated cluster* $X_i \in H_i$, *a cluster* $B \in CP$ *exists such that* $\bigcap_{i=1}^k X_i \subseteq B \subseteq \bigcup_{i=1}^k X_i$.

Faithful consensus rules retain information from a profile P even when only rough agreement exists among its hierarchies. Since $\cap_{i=1}^k X_i$ is the complete agreement (with respect to the X_i) of elements to be clustered in the consensus, betweenness requires some cluster $B \in CP$ to exhibit at least that agreement. Since $\cup_{i=1}^k X_i$ lists the candidates (with respect to the X_i) for clustering in the consensus, betweenness requires that membership in cluster B be restricted to that list. Because of Corollary 4.10 on page 57 and Theorem 4.32 on page 65, quota rules such as *Maj* are not faithful.

Example 4.25. Let $P = (H_1, H_2)$ with $S = abcde$, $H_1^* = \{ab, abc, abcd\}$, and $H_2^* = \{be, bce, bcde\}$ (Figure 4.2 on page 61). If *Maj* were faithful, then *Btw* would hold with $X_1 = abc$ and $X_2 = bce$, so $B \in Maj P^*$ would exist with $bc \subseteq B \subseteq abce$; but $Maj P^* = H_1^* \cap H_2^* = \emptyset$. ■

Yet C_a is faithful.

Lemma 4.26. *The Adams consensus rule* $C_a : \mathcal{H}^k \longrightarrow \mathcal{H}$ *satisfies Btw.*

Proof. Let $P = (H_1, \ldots, H_k) \in \mathcal{H}^k$ and choose clusters $X_i \in H_i$ for all $i \in K$; we will prove that a cluster $B \in C_a P$ exists such that $\cap_{i=1}^k X_i \subseteq B \subseteq \cup_{i=1}^k X_i$. If $(\exists i \in K)(X_i = S)$, the requirement of axiom *Btw* is satisfied for $B = S \in C_a P$. Otherwise let $A = \cap_{i=1}^k X_i$ with $1 < |A|$ to avoid triviality. For each $i \in K$ let $X_i' \in H_i$ be a maximal cluster with $A \subseteq X_i' \subset S$, which must exist since $A \subseteq X_i \subset S$. Consider $B = \cap_{i=1}^k X_i'$. Then $B \in \max C_a P$ in the initialization phase of Algorithm 4.1 on page 60; and if $B \setminus (\cup_{i=1}^k X_i) \neq \emptyset$, then none of the X_i will be deleted from the corresponding H_i. Therefore let Algorithm 4.1 iterate until for the first time $B \setminus (\cup_{i=1}^k X_i) = \emptyset$; the algorithm then assigns B to $C_a P$, and since $\cap_{i=1}^k X_i = A \subseteq B \subseteq \cup_{i=1}^k X_i$, the requirement of axiom *Btw* is satisfied. □

To appreciate faithfulness, consider how betweenness relates to other axioms.

Theorem 4.27. [314, p. 275]. *Let C be a consensus rule on $\mathcal{H}$. C satisfies PO if it satisfies Btw.*

Proof. Let C satisfy *Btw*, let $P = (H_1, \ldots, H_k) \in \mathcal{H}^k$, and let X be such that $X \in H_K$. Use *Btw* with $(\forall i \in K)(X_i = X)$ to see that some $B \in CP$ satisfies $X = \cap_{i=1}^k X_i \subseteq B \subseteq \cup_{i=1}^k X_i = X$, whence *PO* holds. □

Theorem 4.28. [314, p. 276]. *Let C be a consensus rule on $\mathcal{H}$. C satisfies Btw if it satisfies Ind and PO.*

Proof. Let C satisfy *Ind* and *PO*, $P = (H_1, \ldots, H_k) \in \mathcal{H}^k$, and choose $(\forall i \in K)(X_i \in H_i)$. We may assume that $1 < |\cap_{i=1}^k X_i| \leq |\cup_{i=1}^k X_i| < S$, else *Btw* holds trivially. If $\cap_{i=1}^k X_i = \cup_{i=1}^k X_i$, then *PO* implies that $B = \cap_{i=1}^k X_i \in CP$, whence *Btw* holds. Thus let $L = \cap_{i=1}^k X_i$, $U = \cup_{i=1}^k X_i$, $Z = U \setminus L$, and $X = S \setminus Z$. Since $X \subset S$, let $P' = P|_X$. Now $(\forall i \in K)(L = X_i \setminus Z = X_i \cap X \in H_i|_X)$, whence $L \in CP'$ by *PO*. Since $P'|_X = P' = P|_X$, then $CP|_X = CP'|_X$ by *Ind*. Since $L \in CP'$, then $L \in CP'|_X$, whence $L \in CP|_X$. Thus (as can be seen by drawing a Venn diagram) $B \in CP$ exists such that $B \cap X = L$, so $L \subseteq B \subseteq U$, whence *Btw* holds. □

Btw is strictly weaker than *Ind* for rules satisfying *PO*. To see this, associate with each cluster of a hierarchy a measure of that cluster's height in the hierarchy.

Definition 4.29. *Let N be the set of all nonnegative integers. For all $H \in \mathcal{H}$ a* generalized height function *is a function $\eta : H \longrightarrow N$ such that $\eta(S) = 0$ and $\eta(X) > \eta(Y)$ if X and Y are clusters in H with $X \subset Y$. For all $X \in H$, $\eta(X)$ is called its* height *in H. For all $H \in \mathcal{H}$ the* canonical height function *η_0 has, for all $X \in H$, $\eta_0(X) = h$ if and only if there is a sequence $S = X_0 \supset X_1 \supset \cdots \supset X_h = X$ of maximal length with each $X_i \in H$.*

For each $P \in \mathcal{H}^k$ the durchschnitt rule C_d includes in $C_d P$ all clusters formed by intersecting any set of k clusters at the same canonical height, one from each H_i.

Definition 4.30. *The* durchschnitt *consensus rule $C_d : \mathcal{H}^k \longrightarrow \mathcal{H}$ is such that for all $P = (H_1, \ldots, H_k) \in \mathcal{H}^k$ with $\omega = \min_{i \in K} \max_{X \in H_i} \eta_0(X)$,*

$$C_d^* P = \bigcup_{j=1}^{\omega} \{X_1 \cap \cdots \cap X_k : (\forall i \in K)(X_i \in H_i \wedge \eta_0(X_i) = j)\}^*.$$

Although C_d is faithful and thus satisfies *Btw* and *PO*, C_d need not satisfy *Ind*.

Example 4.31. [314, p. 286]. Let $S = abcdefg$ and $X = abcde$. Consider profiles $P = (H_1, H_2) \in \mathcal{H}^2$ and $P' = (H_3, H_2) \in \mathcal{H}^2$, where $H_1^* = \{abf, abcdfg\}$, $H_2^* = \{abce, abcde\}$, and $H_3^* = \{ab, abf, abcdf, abcdfg\}$. Then $C_d^* P = \{ab, abcd\}$ and $C_d^* P' = \{abc, abcd\}$. Although $P|_X = P'|_X$, yet $C_d^* P|_X = C_d^* P \neq C_d^* P' = C_d^* P'|_X$, whence C_d violates *Ind*. ■

With Theorem 4.9 on page 56 the next result, an impossibility theorem, explains for counting rules on hierarchies why only dictatorial rules are faithful.

Theorem 4.32. Neumann [314, p. 286]. *Let C be a consensus rule on* $\mathcal{H}$. *C satisfies Prj if it satisfies DN and Btw.*

Proof. Let C satisfy *DN* but not be a projection; then we will prove that C violates *Btw*. By Theorem 4.8 on page 56 a semidecisive family $\mathcal{S} \subseteq 2^K$ exists such that $C = M_{\mathcal{S}}$. Let $I \in \mathcal{S}$ be a set of minimum cardinality. Since C is not strictly dictatorial, then $|I| > 1$, so let $j \in I$ and construct $P \in \mathcal{H}^k$ such that $H_j^* = \{ab\}$, $(\forall i \in I \setminus \{j\})(H_i^* = \{abc\})$ and $(\forall i \in K \setminus I)(H_i^* = \{abd\})$. Since $\{j\}$, $I \setminus \{j\}$, and $K \setminus I$ are not in $\mathcal{S}$, then $C^* P = M_{\mathcal{S}}^* P = \emptyset$. Now apply *Btw* to P. For each $H_i \in P$ let X_i be the unique nontrivial cluster, whence $\cap_{i=1}^k X_i = ab$ and $\cup_{i=1}^k X_i = abcd \neq S$; but since $C^* P = \emptyset$, C violates *Btw* on P. □

Open Problem 4.33. Characterize consensus rules on hierarchies that satisfy *Btw*. ■

4.2.3 Generalized Intersection Rules

The concept of cluster height arises naturally (Figure 4.3) when using hierarchies to analyze data. It occurs explicitly when defining the durchschnitt rule, consensus clusters being set intersections of profile clusters at the same height with respect to the canonical height function η_0. It occurs implicitly when specifying the Adams consensus rule, where on any given iteration Algorithm 4.1 on page 60 identifies in each current profile hierarchy a set of nontrivial maximal clusters with respect to set inclusion: at that point such clusters are of canonical height one. Cluster cardinality also yields a natural measure of cluster height.

Definition 4.34. *For all* $H \in \mathcal{H}$ *on* S, *the* cardinality height function *is a function* $\eta_\sharp : H \longrightarrow N$ *such that* $(\forall X \in H)(\eta_\sharp(X) = |S| - |X|)$.

Pairing hierarchy with height function yields a data structure called variously a ranked tree, valued tree, dendrogram [184, p. 68], or numerically stratified clustering [212, p. 61]; it fixes the relative positions of a hierarchy's nontrivial clusters so as to make that information available for data analyses.

Example 4.35. Consider hierarchy H on $S = abcde$ with $H^* = \{ab, cde, de\}$. Figure 4.4 on page 67 shows H with four height functions: h_0 is the canonical height function, so $h_0(ab) = h_0(cde) = 1$ and $h_0(de) = 2$; $h_\sharp$ is the cardinality height function, so $h_\sharp(ab) =$

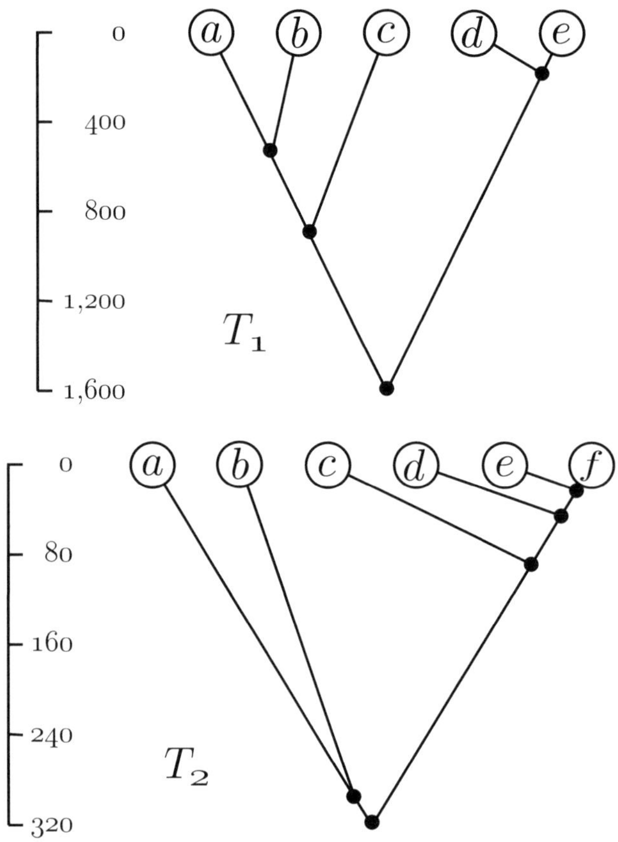

Figure 4.3. *Cluster Heights in Hierarchies. Associated with each interior vertex is a time from the vertical scale, in millions of years before present, when that speciation event may have occurred. T_1 depicts the evolutionary history of multicellular organisms whose genomes have been sequenced* [64, p. 61]. *a: human (mammal), b: fruit fly (arthropod), c:* Caenorhabditis elegans *(nematode), d:* Arabidopsis *(dicotyledonous plant), e: rice (monocotyledonous plant). T_2 depicts the evolutionary history of ant wing polyphenism* [1, p. 252]. *a: Drosophila, b: Precis, c: Neoformica, d: Myrmica, e: Crematogaster, f: Pheidole.*

$h_\sharp(de) = 3$ and $h_\sharp(cde) = 2$; h_t is a time function where $h_t(ab) < h_t(cde)$, i.e., the speciation of a and b occurred before the speciation of c and a precursor of d and e; h_s is a similarity function where d is more similar to e than a is to b. ■

To exploit the relative positions of clusters in hierarchies, let a generalized intersection consensus rule be defined such that a generalized height function (Definition 4.29 on page 64) specifies each cluster's height, and each consensus cluster is the set intersection of k clusters, one in each profile hierarchy and all essentially at the same height. Since a height function is defined for one hierarchy, extend the concept to all hierarchies of a profile: a *height*

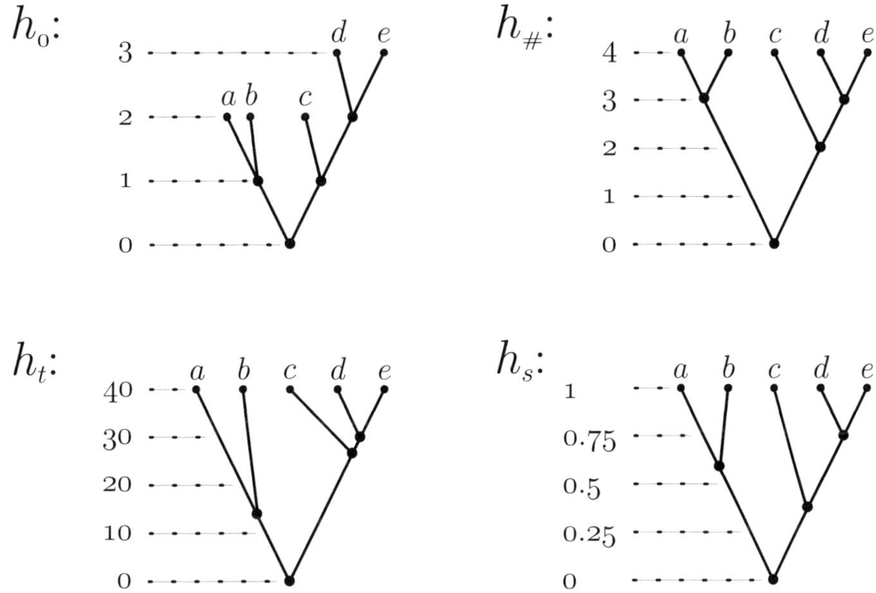

Figure 4.4. *Hierarchies with Height Functions* I. *Example* 4.35 *explains.*

assignment on $\mathcal{H}^k$ is a function h on $\mathcal{H}^k$ that assigns to each $P = (H_1, \ldots, H_k)$ a k-tuple $(\eta_1, \ldots, \eta_k)$, where each $\eta_i = \eta_i(P)$ is a generalized height function on H_i. Given $H_i \in P$, $x \in S$, and $r \in N$, identify (if it exists) a cluster of H_i that contains x and is of height at least r: for all $i \in K$ let $F_i(x, P, r)$ be the coarsest cluster in H_i that contains x and whose height is at least r, the convention being that $F_i(x, P, r) = \emptyset$ when no such cluster exists. Since at each height we wish to take the set intersection of all such profile clusters, let $F(x, P, r) = \cap_{i=1}^{k} F_i(x, P, r)$ and consider this definition.

Definition 4.36. *For each height assignment h let $C_h : \mathcal{H}^k \longrightarrow \mathcal{H}$ be the* generalized intersection consensus rule *such that*

$$(\forall P \in \mathcal{H}^k)\left(C_h^* P = \bigcup_{r \in \mathbf{N}} \{F(x, P, r) : x \in S\}^*\right).$$

The durchschnitt rule (*Definition* 4.30) *is the rule $C_d = C_h$, where h assigns the canonical height function η_0 to every profile hierarchy. The* cardinality intersection rule *is the rule $C_\sharp = C_h$, where h assigns the cardinality height function $\eta_\sharp$ to every profile hierarchy.*

Example 4.37. In Figure 4.5 consider profile $P = (H_1, H_2)$ on $S = abcdefg$ with $H_1^* = \{ab, abc, abcd, abcde, abcdef\}$ and $H_2^* = \{ab, abcd, cd, efg\}$. If $h_0 = (\eta_0, \eta_0)$, then $C_d^* P = C_{h_0}^* P = \{ab, cd, abcd, ef\}$. If $h_\sharp = (\eta_\sharp, \eta_\sharp)$, then $C_\sharp^* P = C_{h_\sharp}^* P = \{ab, abcd, ef\}$. ■

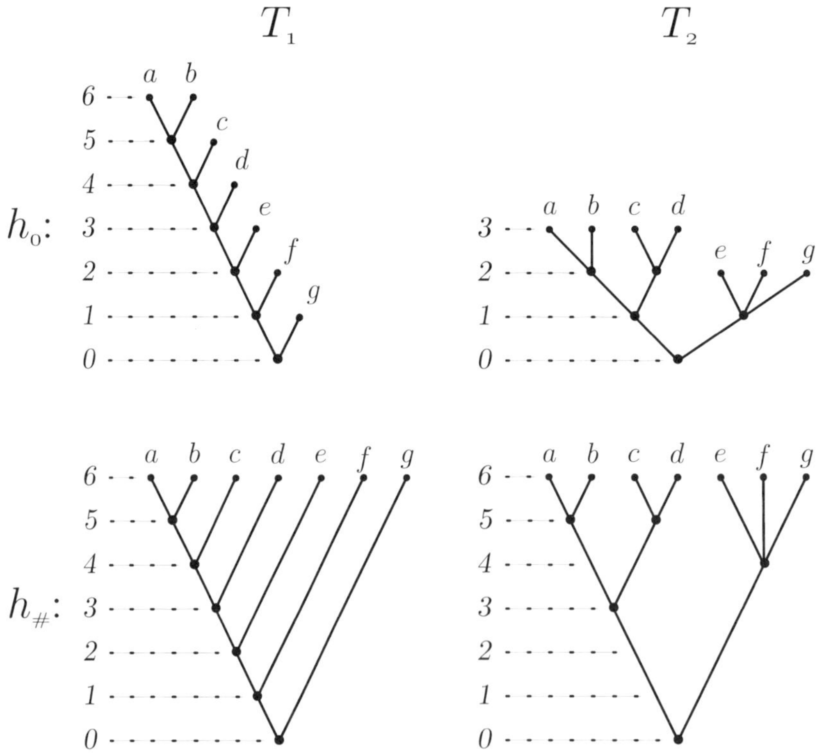

Figure 4.5. *Hierarchies with Height Functions* II. *Column i shows tree representations of hierarchy H_i. Row i applies height assignment $h_i = (\eta_i, \eta_i)$ to the tree representations. Example* 4.37 *explains.*

Powers's Characterizations

To characterize the C_h rules Powers [334, p. 52] develops a modification of betweenness. Let $P = (H_1, \ldots, H_k) \in \mathcal{H}^k$ be a profile with height assignment $h = (\eta_1, \ldots, \eta_k)$. For each $i \in K$, in H_i there is at least one cluster $X \neq S$ that is maximal with respect to subset inclusion. Among them at least one is at a minimal height, $m_i(P, h)$, where

$$(\forall i \in K)(m_i(P, h) = \min\{\eta_i(X) : X \in H_i \wedge X \neq S\}).$$

The minimum such height, taken over all $i \in K$, is

$$m(P, h) = \min\{m_1(P, h), \ldots, m_k(P, h)\}.$$

Among the nontrivial clusters in each $H_i \in P$, some may be at this minimum height:

$$(\forall i \in K)(U_i(P, h) = \{X \in H_i^* : \eta_i(X) = m(P, h)\}).$$

Consider the profile obtained by removing from each $H_i \in P$ the clusters of $U_i(P, h)$:

$$P \setminus U(P, h) = (H_1 \setminus U_1(P, h), \ldots, H_k \setminus U_k(P, h)). \tag{4.4}$$

Any cluster heights in $P \setminus U(P, h)$ should be reasonably related to those in P as in the following definition.

Definition 4.38. *For all $P = (H_1, \ldots, H_k) \in \mathcal{H}^k$ with $h = (\eta_1, \ldots, \eta_k)$, let $h(P \setminus U(P, h)) = (\eta'_1, \ldots, \eta'_k)$. The height assignment h is called* standard *if*

$$\eta_i(X) > \eta_j(Y) \Longleftrightarrow \eta'_i(X) > \eta'_j(Y)$$

for all $i, j \in K$ with $X \in H_i \setminus U_i(P, h)$ and $Y \in H_j \setminus U_j(P, h)$. For example, the height assignment $h = (\eta_1, \ldots, \eta_k) = (\eta_0, \ldots, \eta_0) = h_0$ is standard.

How should $C(P \setminus U(P, h))$ and CP be related? Since in $P \setminus U(P, h)$ we merely deleted clusters at the lowest possible height, one might expect that any clusters in $C(P \setminus U(P, h))$ at higher levels should be in CP, i.e., that $C(P \setminus U(P, h)) \subseteq CP$. Similarly, if $\max H$ is the set of all nontrivial maximal clusters of H with respect to set inclusion, one might expect that $CP \setminus \max CP \subseteq C(P \setminus U(P, h))$. Such are the motivations for a new concept of betweenness.

Definition 4.39. *If h is a height assignment on $\mathcal{H}^k$, then consensus rule $C : \mathcal{H}^k \longrightarrow \mathcal{H}$ satisfies h-*betweenness *(Btw_h) if*

$$(\forall P \in \mathcal{H}^k)(CP \setminus \max CP \subseteq C(P \setminus U(P, h)) \subseteq CP),$$

$P \setminus U(P, h)$ being defined by (4.4).

By weakening qualified strong presence (*QSP*) to upper strong presence (*USP*), the generalized intersection rules can be characterized.

Theorem 4.40. Powers [334, p. 53]. *Let h be a standard height assignment on $\mathcal{H}^k$ with C a consensus rule on $\mathcal{H}$. C satisfies NP, USP, and Btw_h if and only if $C = C_h$.*

Readers may consult [334, pp. 53–54] for a proof. Moreover the durchschnitt rule is characterized by using the canonical height assignment in Theorem 4.40.

Corollary 4.41. *Let $h_0 = (\eta_0, \ldots, \eta_0)$ be the canonical height assignment on $\mathcal{H}^k$ with C a consensus rule on $\mathcal{H}$. C satisfies NP, USP, and Btw_{h_0} if and only if $C = C_d$.*

By making suitable changes to Theorem 4.40, Powers also characterizes the Adams consensus rule. Now the role of the $U_i(P, h)$ is played by the $V_i(P)$ in (4.2b), and $P \setminus U(P, h)$ is replaced by $P \setminus V(P)$ in (4.2c). How should $C(P \setminus V(P))$ and CP be related? Since in $P \setminus V(P)$ we merely deleted clusters at the lowest possible level, one might expect that any clusters in $C(P \setminus V(P))$ at higher levels should be in CP, i.e., that $C(P \setminus V(P)) \subseteq CP$. Similarly, if $\max H$ is the set of all nontrivial maximal clusters of H with respect to set inclusion, one might expect that $CP \setminus \max CP \subseteq C(P \setminus V(P))$. Such are the motivations for a new concept of betweenness, one that can be used to characterize C_a.

Definition 4.42. *Consensus rule* $C : \mathcal{H}^k \longrightarrow \mathcal{H}$ *satisfies* a-betweenness (Btw_a) *if*

$$(\forall P \in \mathcal{H}^k)(CP \setminus \max CP \subseteq C(P \setminus V(P)) \subseteq CP),$$

$P \setminus V(P)$ *being defined by* (4.2c).

Theorem 4.43. Powers [334, p. 55]. *Let* C *be a consensus rule on* $\mathcal{H}$. C *satisfies NP, USP, and* Btw_a *if and only if* $C = C_a$.

4.3 Median Rules

Where before consensus rules had the form $C : \mathcal{X}^k \longrightarrow \mathcal{X}$, now we will study complete multiconsensus rules. They can be based on a measure of remoteness of an object to a profile [54, p. 241]: sum the distances from an object $X \in \mathcal{X}$ to those in a profile P; if the value of that measure is minimum on $\mathcal{X}$, call X a *median* and make it a consensus object for P. Consider such concepts specifically for hierarchies.

Definition 4.44. *For* $\mathcal{H}$ *the set of all hierarchies on* S, *the* symmetric difference distance *between any* $H, H' \in \mathcal{H}$ *is the number*

$$d(H, H') = |H \cup H'| - |H \cap H'| = |H| + |H'| - 2|H \cap H'|$$

of clusters in one hierarchy but not both. A median *of* $P = (H_1, \ldots, H_k) \in \mathcal{H}^k$ *is any* $H \in \mathcal{H}$ *for which the* remoteness $r(H, P) = \sum_{i=1}^{k} d(H, H_i)$ *is minimum on* $\mathcal{H}$. *The method of* median rule *on* $\mathcal{H}$ *is a rule* $Med : \mathcal{H}^* \longrightarrow 2^{\mathcal{H}} \setminus \{\emptyset\}$ *such that*

$$(\forall P \in \mathcal{H}^k)(Med(P) = Med\,P = \{H \in \mathcal{H} : H \text{ is a median of } P\}).$$

On $\mathcal{H}$ the majority (Definition 3.47 on page 44) and median rules are closely related.

Lemma 4.45. [246, p. 242]. *If* $P \in \mathcal{H}^k$, *then* $Maj\,P \in Med\,P$, *and* $Med\,P = \{Maj\,P\}$ *when* k *is odd.*

Proof. For each $H \in \mathcal{H}$ let $I(H)$ be an incidence vector, of length $2^{|S|}$, where

$$I(H)_j = \begin{cases} 1 & \text{if the cluster identified by } j \text{ is in } H, \\ 0 & \text{otherwise.} \end{cases}$$

Let $P = (H_1, \ldots, H_k) \in \mathcal{H}^k$ with k odd, $H \in \mathcal{H}$, and $H' = Maj\,P$. Then

$$\sum_{i \in K} d(H', H_i) = \sum_{i \in K} \sum_{j} |I(H')_j - I(H_i)_j|$$

and

$$\sum_{i \in K} d(H, H_i) = \sum_{i \in K} \sum_j |I(H)_j - I(H_i)_j|.$$

If $I(H)_j \neq I(H')_j$, then

$$\sum_{i \in K} |I(H)_j - I(H_i)_j| > \sum_{i \in K} |I(H')_j - I(H_i)_j| \tag{4.5a}$$

since $I(H')_j$ and $I(H_i)_j$ are the same for a majority of the H_i in P. Thus for all j

$$\sum_{i \in K} |I(H)_j - I(H_i)_j| \geq \sum_{i \in K} |I(H')_j - I(H_i)_j|,$$

and the inequality is strict if $I(H)_j \neq I(H')_j$. Then, summing over j,

$$\sum_{i \in K} d(H, H_i) > \sum_{i \in K} d(H', H_i) \tag{4.5b}$$

unless $H = H'$, whence H' is the unique median for P. For k even, the inequalities at (4.5a) and (4.5b) need not be strict, so medians need not be unique. □

Example 4.46. Let H^* be the set of all nontrivial clusters of $H \in \mathcal{H}$ and let $S = abc$. If $P = (H_1, H_2, H_3)$ and $P^* = (H_1^*, H_2^*, H_3^*) = (\{ab\}, \{ac\}, \{ac\})$, then

H^* :	$\emptyset$	$\{ab\}$	$\{ac\}$	$\{bc\}$
$r(H, P)$:	3	4	2	6

so $Maj P = H_2 = H_3$ and $Med P = \{Maj P\}$. If $P = (H_1, H_2)$ and $P^* = (H_1^*, H_2^*) = (\{ab\}, \{ac\})$, then

H^* :	$\emptyset$	$\{ab\}$	$\{ac\}$	$\{bc\}$
$r(H, P)$:	2	2	2	4

so $Maj P = H_\emptyset$ and $Med P = \{Maj P, H_1, H_2\}$. ■

Excepting the majority rule hierarchy, what is in the median set? Our answer uses the cluster index γ (Definition 3.46 on page 44) and a new concept of compatibility.

Definition 4.47. *A nonempty set* $A \subseteq S$ *is called* compatible with $H \in \mathcal{H}$ *if* $A \cap X \in \{\emptyset, A, X\}$ *for all* $X \in H^*$, *i.e.,* A *is compatible with* H *if* $H \cup \{A\}$ *remains a hierarchy.*

In a sense the majority rule hierarchy is the nucleus of every median hierarchy.

Lemma 4.48. [49, p. 332]. *For all* $P \in \mathcal{H}^k$, $Med P$ *is the set of all hierarchies* $Maj P \cup \{A_1, \ldots, A_m\}$ *such that for* $1 \leq l \leq m$, $\gamma(A_l, P) = \frac{1}{2}$ *and* A_l *is compatible with* $Maj P \cup \{A_1, \ldots, A_{l-1}\}$.

Proof. Let $P = (H_1, \ldots, H_k) \in \mathcal{H}^k$. If k is odd, then $Med\,P = \{Maj\,P\}$ by Lemma 4.45 on page 70. Let k be even and use induction on m. For the basis step with $m = 1$, let $H = Maj\,P$, $\gamma(A_1, P) = \frac{1}{2}$, A_1 be compatible with P, and set $H' = H \cup \{A_1\}$. We must show that $\sum_{i=1}^{k} d(H', H_i) = \sum_{i=1}^{k} d(H, H_i)$. Using the distance definition,

$$\begin{aligned}\sum_{i=1}^{k} d(H', H_i) &= k|H'| + \sum_{i=1}^{k} |H_i| - 2\sum_{i=1}^{k} |H' \cap H_i| \\ &= k(|H| + 1) + \sum_{i=1}^{k} |H_i| - 2\left(\sum_{i=1}^{k} |H \cap H_i| + \frac{k}{2}\right) \\ &= k|H| + \sum_{i=1}^{k} |H_i| - 2\sum_{i=1}^{k} |H \cap H_i| \\ &= \sum_{i=1}^{k} d(H, H_i).\end{aligned}$$

The inductive step is similar. □

But the median set may be unacceptably large.

Example 4.49. Let $S = abcd$. If $P = (H_1, H_2) = (H_\emptyset \cup \{ab, cd\}, H_\emptyset \cup \{ac, bd\})$, then $Maj\,P = H_\emptyset$. By Lemma 4.48 the median set has all hierarchies whose nontrivial clusters are subsets of $\{ab, cd\}$ or $\{ac, bd\}$, so

$$Med\,P = \{H \in \mathcal{H} : H^* \in \{\emptyset, \{ab\}, \{cd\}, \{ab, cd\}, \{ac\}, \{bd\}, \{ac, bd\}\}\}.$$

This is the smallest example in a family for which $|Med\,P| = 2^{\frac{n}{2}+1} - 1$, where $n = |S|$. In the worst case the number of median hierarchies increases exponentially with n, even though calculating $Maj\,P \in Med\,P$ requires time at most polynomial in n. ■

Consider what axioms would be suitable to describe consensus rules of the form $C : \mathcal{H}^* \longrightarrow 2^{\mathcal{H}} \setminus \{\emptyset\}$. Our candidates will use these concepts and notations.

Definition 4.50. *$X \subseteq S$ is called a C-*solution cluster* of $P \in \mathcal{H}^*$ if $X \in H$ for some $H \in CP$; $Q(C, P)$ is the set of all C-solution clusters of P. For all $\emptyset \neq X \subseteq S$ let $H_X = H_\emptyset \cup \{X\}$. For all $H \in \mathcal{H}$ let $H^X = H_X$ if $X \in H$, or $H^X = H_\emptyset$ if $X \notin H$. For all $P = (H_1, \ldots, H_k) \in \mathcal{H}^*$ let $P^X = (H_1^X, \ldots, H_k^X)$. Finally PP' is the concatenation of profiles P and P'.*

In Table 4.3 on the facing page independence (*Ind*) is a form of decisive neutrality (Table 4.1 on page 54): if a cluster X is in hierarchies at the same positions of profiles P and P', then it is a C-solution cluster of P if and only if it is a C-solution cluster of P'. Efficiency (*Eff*) restricts the consensus result of profiles in two situations: $H \in \mathcal{H}$ must be the unique median hierarchy when H is at every position of a profile; for profiles in which only H_X and $H_\emptyset$ occur, the consensus result must be restricted to those two hierarchies. For

Table 4.3. *Axioms: Rules on Hierarchies* IV. *For notation see Definitions* 3.49 (*on page* 44) *and* 4.50.

***Cnd*: Condorcet** $(\forall X \subseteq S)(\forall P \in \mathcal{H}^*)(\forall H \in \mathcal{H})$
$(\gamma(X, P) = \frac{1}{2} \wedge X$ compatible with $H \Longrightarrow [H \in CP \Longleftrightarrow H \cup \{X\} \in CP])$

***Css*: Consistency** $(\forall P, P' \in \mathcal{H}^*)(CP \cap CP' \neq \emptyset \Longrightarrow C(PP') = CP \cap CP')$

***Eff*: Efficiency** (i) $(\forall H \in \mathcal{H})(C((H)_k) = \{H\})$;
(ii) $(\forall X \subseteq S)(\forall P \in \mathcal{H}^k)(C(P^X) \subseteq \{H_X, H_\emptyset\})$

***Fth*: Faithfulness** $(\forall P = (H_1) \in \mathcal{H}^1)(CP = \{H_1\})$

***Ind*: Independence** $(\forall X \subseteq S)(\forall P, P' \in \mathcal{H}^k)$
$(K_X(P) = K_X(P') \Longrightarrow [X \in Q(C, P) \Longleftrightarrow X \in Q(C, P')])$

***Opt*: Optimality** $(\forall P \in \mathcal{W}^*)(CP \subseteq \{W \in \mathcal{W} : \sigma_P(W)$ is minimum$\})$, where
$J(P) = \left\{X \subseteq S : \gamma(X, P) \geq \frac{1}{2}\right\}$ and

$$\sigma_P(W) = \sum_{X \in J(P) \setminus W} k(2\gamma(X, P) - 1)$$

***Sym*: Symmetry**
$(\forall P \in \mathcal{H}^k)(\forall$ permutations σ of $K)(Q(C, P) = Q(C, (H_{\sigma(1)}, \ldots, H_{\sigma(k)})))$

consistency (*Css*) let a society's members meet in two rooms, with each room using rule C to elect Jones as an officer: since the two rooms separately elect Jones, should not the combined rooms using rule C also elect Jones? Condorcet (*Cnd*) generalizes the feature of median sets described by Lemma 4.48 on page 71: if a cluster A is in half the hierarchies of profile P and is compatible with H, then $H \in CP$ if and only if $H \cup \{A\} \in CP$. Its name honors the Marquis de Condorcet (1743–94), a social philosopher, mathematician, and political leader who died in prison during the French revolution. Such axioms characterize the median rule for hierarchies.

Theorem 4.51. Barthélemy and McMorris [49, p. 333]. *Let* $C : \mathcal{H}^* \longrightarrow 2^{\mathcal{H}} \setminus \{\emptyset\}$ *be a consensus rule.* $C = Med$ *if and only if* C *satisfies Cnd, Css, Eff, Ind, and Sym.*

Proof. Using Lemma 4.48 on page 71 and the definitions, it is easy to see that *Med* satisfies the five axioms. Thus let C satisfy the five axioms. Because of *Cnd* and Lemma 4.48, it will suffice to prove for each $X \subseteq S$, $P \in \mathcal{H}^k$, and $H \in CP$ that $(\gamma(X, P) > \frac{1}{2} \Longrightarrow X \in H)$ and $(\gamma(X, P) < \frac{1}{2} \Longrightarrow X \notin H)$.

Let $\gamma(X, P) > \frac{1}{2}$ and suppose X is a cluster in m hierarchies of P, where $m > k - m$. Let $P_1 \in \mathcal{H}^{2m-k}$ with $P_1 = (H, \ldots, H)$ and set $P_2 = PP_1$. Then $CP_1 = \{H\}$ by *Eff*, and $CP_2 = \{H\}$ by *Css*. Now $C(P_2^X) \subseteq \{H_X, H_\emptyset\}$ by *Eff*, so X is compatible with every hierarchy in $C(P_2^X)$. Thus $X \in Q(C, P_2^X)$ by *Cnd*, so that $X \in Q(C, P_2)$ by *Ind*, whence $X \in H$.

Let $\gamma(X, P) < \frac{1}{2}$ and suppose $X \in Q(C, P)$ so that $X \in Q(C, P^X)$ by *Ind*. Suppose H_X occurs m times in P^X, where $m < k - m$. Form P' from P^X by deleting $k - 2m$ occurrences of $H_\emptyset$, so that H_X and $H_\emptyset$ occur equally often in P'; by *Sym* it does not matter which occurrences of $H_\emptyset$ are deleted. Let $P_\emptyset = (H_\emptyset, \ldots, H_\emptyset) \in \mathcal{H}^{k-2m}$, so that $CP_\emptyset = \{H_\emptyset\}$ by *Eff*. Then $CP' = \{H_\emptyset, H_X\}$ by *Cnd* and *Eff*. Since $CP' \cap CP_\emptyset = \{H_\emptyset\}$, then $C(P'P_\emptyset) = \{H_\emptyset\}$ by *Css*. Since $Q(C, P^X) = Q(C, P'P_\emptyset)$ by *Sym*, then $X \notin Q(C, P^X)$, a contradiction since $X \in Q(C, P^X)$. □

Another characterization not only avoids using *Ind* and *Sym*, it substitutes for *Eff* the considerably stronger faithfulness axiom, *Fth*, which requires of every $H \in \mathcal{H}$ that it be the unique median hierarchy of the profile $P = (H) \in \mathcal{H}^1$.

Theorem 4.52. McMorris, Mulder, and Powers [264, p. 229]. *Let $C : \mathcal{H}^* \longrightarrow 2^{\mathcal{H}} \setminus \{\emptyset\}$ be a consensus rule. $C = Med$ if and only if C satisfies Cnd, Css, and Fth.*

This result follows from a characterization of the median rule on a median semilattice (Theorem 5.44 on page 97), which follows in turn from a characterization of the median rule on a median graph (Theorem 5.35 on page 93). But in contrast to the situation for hierarchies, complete multiconsensus rules on weak hierarchies exist that satisfy *Cnd*, *Css*, and *Fth* yet are not the median rule.

Example 4.53. [270, p. 514] Let $c : \mathcal{W}^* \longrightarrow 2^{\mathcal{W}} \setminus \{\emptyset\}$ be the rule such that $(\forall P \in \mathcal{W}^k)(cP = \{H \in \mathcal{W} : R(H, P) \text{ is maximum}\})$, remoteness R being such that

$$(\forall H \in \mathcal{W})(\forall P \in \mathcal{W}^k)(R(H, P) = \sum_{A \in H^*} |A|(2k\gamma(A, P) - k)).$$

Readers may consult [270] to confirm that c satisfies *Cnd*, *Css*, and *Fth*. An example on $S = abcdefgh$ shows that $c \neq Med$: if $P = (H_1, H_2, H_3) \in \mathcal{W}^3$, where $H_1^* = \{abcdef, ag, agh\}$, $H_2^* = \{abcdef, fg, fgh\}$, and $H_3^* = \{ag, agh, fg, fgh\}$, then $cP = \{H_1, H_2\} \neq \{H_3\} = Med\,P$. ■

To characterize consensus rules on weak hierarchies, McMorris and Powers impose a pointwise ordering, i.e., $C_1 \leq C_2$ if and only if $C_1(P) \subseteq C_2(P)$ for all $P \in \mathcal{W}^k$, and in the optimality (*Opt*) axiom they score weak hierarchies relative to any given profile.

Theorem 4.54. [272, p. 268]. *The median rule on $\mathcal{W}$ is the maximum element in the set of all rules $C : \mathcal{W}^* \longrightarrow 2^{\mathcal{W}} \setminus \{\emptyset\}$ that satisfy Cnd, Css, Fth, and Opt.*

Readers may find that result satisfying only in a technical sense.

Open Problem 4.55. Use conceptually simple properties to characterize the median rule on $\mathcal{W}$. ■

Open Problem 4.56. [272, p. 268]. Characterize the set of complete multiconsensus rules on $\mathcal{W}$ that satisfy *Cnd*, *Css*, and *Fth*. ■

4.4 Notes

Mickevich's [279, 280] use of consensus hierarchies to summarize areas of agreement among hierarchies stimulated the development of consensus rules for classification and systematics in the 1980s. Popular rules for hierarchies include the Adams consensus rule (Adams [2, 3]), the majority consensus rule (Margush and McMorris [246]), and the *strict* consensus rule (Sokal and Rohlf [383, p. 312], Day [149]). The *combinable component* (Bremer [109]), *semistrict* or *loose* (Meacham [276], Barthélemy, McMorris, and Powers [52]) consensus rule is based on cluster compatibility (Definition 4.47 on page 71): the loose consensus of a profile $P = (T_1, \ldots, T_k)$ of hierarchies contains exactly those clusters of $\cup_{i=1}^{k} T_i$ that are compatible with every hierarchy in P. Swofford [399] nicely summarizes various elucidations (Page [319], Bremer [109]) of Nelson's [313] consensus rule. Wilkinson [413, 415] addresses problems of insensitivity and ambiguity in the strict and Adams consensus rules by developing new consensus rules based on obtaining reduced subtrees by pruning leaves. Phillips and Warnow [328] describe an *asymmetric median* consensus rule for hierarchies, its consensus results being at least as informative as majority-rule consensus hierarchies. Kannan, Warnow, and Yooseph [216] develop a *local* consensus rule for hierarchies, which estimates for every three objects the corresponding triad of the consensus hierarchy. For assessments of consensus rules on hierarchies by theoretical or empirical means see Shao [376], Swofford [399], Swofford et al. [400], Bryant [112, 113], and Page and Holmes [320]. For discussions of appropriate uses of consensus rules in systematic biology see Miyamoto [288], Hillis [205], Barrett, Donoghue, and Sober [38], and Wilkinson [413].

Section 4.1 (Counting Rules): For decisive-family rules on hierarchies see McMorris and Neumann [267]; for decisive-family rules on weak hierarchies see McMorris and Powers [268]; for other papers on quota rules see Mirkin [285], Barthélemy [44], Monjardet [296], and Barthélemy and Janowitz [46].

Section 4.2 (Intersection Rules): Intersection rules are explored by Neumann and Norton [315], Stinebrickner [390, 391, 392], Vach [405], Powers [334], and McMorris and Powers [271].

Section 4.3 (Median Rules): The median rule on hierarchies is investigated by Margush and McMorris [246], Barthélemy and McMorris [49], McMorris and Steel [275], McMorris and Powers [270], and McMorris, Mulder, and Powers [264]; for the median rule on weak hierarchies see McMorris and Powers [272]. Barthélemy and Monjardet [54, 55] review early investigations of the median rule in data analysis and social choice theory, while Monjardet [298] gives a history of the median metric. Barthélemy and Leclerc [47] review the literature on the median rule, particularly with respect to partitions of a set. Barthélemy and Janowitz [46], McMorris and Powers [270], McMorris [261], McMorris, Mulder, and Roberts [266], and McMorris, Mulder, and Powers [264] investigate the median rule in abstract settings involving ordered sets or graphs. McMorris [260, 261] gives a maximum likelihood interpretation of the median rule for hierarchies and for other restricted classes of hypergraphs; Young [425, 426] argues that such likelihood interpretations are natural for the Borda [100] and Condorcet [137] rules. Other papers using the general hypergraph model include McMorris and Powers [271] and Lehel, McMorris, and Powers [241].

Chapter 5

General Models of Consensus

What is needed is a general mathematical model in which [Arrovian] matters may be disposed of in a common setting. That is to say, we forget about the exact nature of the objects and, using some abstract structure on various sets of objects under consideration, concern ourselves instead with ways in which the structure can be used to summarize a given family of objects.
— J. P. Barthélemy and M. F. Janowitz [46, p. 305]

Our impossibility theorems have had essentially the same forms (Template 1.8 on page 7 and Template 3.2 on page 28); our possibility theorems, i.e., characterizations, have used axioms that recur in different contexts. Indeed, general results can explain, or provide useful frameworks in which to study, various concrete results in Chapters 3 and 4. We will describe investigations of Arrovian paradigms: to formalize the aggregation of partitions of a set (section 5.2.1) and consensus rules based on decisive families (section 5.2.2), to extend the median rule on hierarchies to more abstract settings (section 5.3), and to generalize the median rule via the concept of remoteness (section 5.4).

5.1 Ordered Sets

The partially ordered set provides a relevant context for these investigations since with it the consensus of a set of objects is an object that bounds the set in a manner determined by the set's partial order.

Definition 5.1. *A* partially ordered set *or* poset *is an ordered pair* (X, R) *where* X *is a set and* R *is a partial order on* X *(Table* 1.3 *on page* 6*) that is denoted by* $\leq$ *or* $R_{\leq}$*. A finite poset* $(X, \leq)$ *can be depicted by a* diagram $G = (X, E)$*, where* X *is the vertex set,* E *is the set of directed edges, and* $(x, y) \in E$ *if and only if* y covers x*, i.e.,* $x < y$ *and* $x < z < y$ *for no* $z \in X$*. To draw a diagram we place its vertices so that* x *lies below* y *if* $xy \in E$*, so the relative positions of an edge's end points show its orientation. A poset* $(X, \leq)$ *also has*

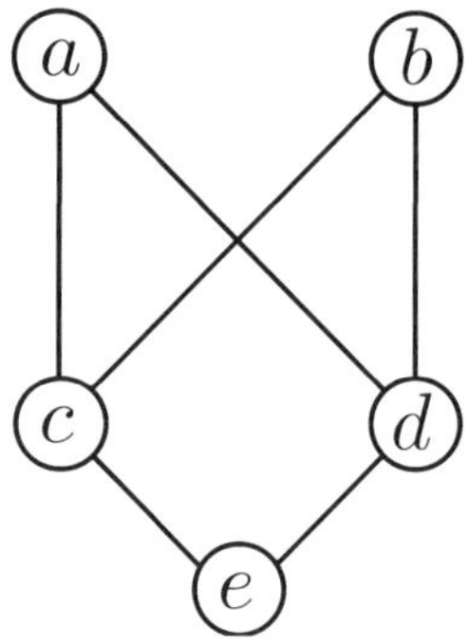

Figure 5.1. *Diagram of the Poset in Example 5.2*

an undirected covering graph $G = (X, E)$*, where* $\{x, y\} \in E$ *if and only if* x *covers* y *or* y *covers* x *in* $(X, \leq)$.

Example 5.2. Consider the poset $(X, R_{\leq})$, where $X = abcde$ and

$$R_{\leq} = \{aa, bb, cc, dd, ee, ca, cb, da, db, ea, eb, ec, ed\}.$$

Figure 5.1 has its diagram $G = (X, E)$, where $E = \{ca, cb, da, db, ec, ed\}$. ■

In a poset $(X, \leq)$ if $a, b \in X$ and $a \leq b$, then a is a lower bound of b and b is an upper bound of a. The concepts generalize naturally as in the following definition.

Definition 5.3. *Let* $Y \subseteq X$ *and* $a \in X$*. Then* a *is a* lower bound *of* Y *if* $a \leq y$ *for all* $y \in Y$*;* a *is an* upper bound *of* Y *if* $a \geq y$ *for all* $y \in Y$*. A lower bound* a *of* Y *is a* greatest lower bound *of* Y *if* $b \leq a$ *for every lower bound* b *of* Y*; an upper bound* a *of* Y *is a* least upper bound *of* Y *if* $b \geq a$ *for every upper bound* b *of* Y.

Upper/lower bounds need not exist or if they do they need not be unique. In Figure 5.1 $Y = \{a, b\}$ has no upper bounds yet c, d, and e are its lower bounds. Least upper and greatest lower bounds need not exist, but if they do they are unique. When they exist we will denote the greatest lower bound or *meet* of Y by $\wedge Y$ and the least upper bound or *join* of Y by $\vee Y$. If $Y = \{a, b\}$, we may write $a \wedge b$ instead of $\wedge Y$ and $a \vee b$ instead of $\vee Y$. In Figure 5.1 $Y = \{c, d\}$ has $e = \wedge Y = c \wedge d$, yet it has no join. But posets do exist in which meet (or join) elements are associated with every set of elements.

Definition 5.4. *A* meet semilattice $L = (L, \leq)$ *is a poset such that* $x \wedge y$ *exists for all* $x, y \in L$*. If* L *is finite,* $\wedge L$ *is called the* least element *or* zero *of* L *and is denoted by* 0_L *or* 0*. A* join semilattice $L = (L, \leq)$ *is a poset such that* $x \vee y$ *exists for all* $x, y \in L$*. If* L *is finite,* $\vee L$ *is called the* greatest element *or* unit *of* L *and is denoted by* 1_L *or* 1*. A* lattice *is a poset that is both a meet and a join semilattice.*

Among properties of lattices distributivity is useful and well known.

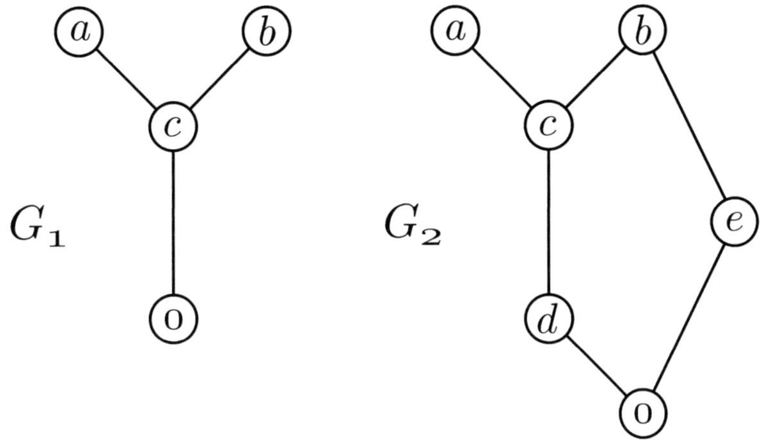

Figure 5.2. *Diagrams G_i of Meet Semilattices L_i. L_1 is lower distributive and L_2 is not. After deleting vertex a and edge ca, the resulting diagrams G_i' depict lattices, where L_1' is distributive and L_2' is not.*

Definition 5.5. *A lattice $L = (L, \leq)$ is called* distributive *if*

$$(\forall x, y, z \in L)((x \vee y) \wedge z = (x \wedge z) \vee (y \wedge z)).$$

Let $L = (L, \leq)$ be a meet semilattice. If $x \in L$, then $(x] = \{y \in L : y \leq x\}$ is always a lattice. A meet semilattice L is called lower distributive *if $(x]$ is distributive for all $x \in L$. Figure 5.2 illustrates the lower distributivity and distributivity concepts.*

Among properties of semilattices the median establishes conditions under which for each $x, y, z \in L$ some element exists that in the covering graph of L is on shortest paths between x and y, x and z, and y and z.

Definition 5.6. *A meet semilattice L satisfies the* median *or* join-Helly *property if for all $x, y, z \in L$ such that $x \vee y$, $x \vee z$, and $y \vee z$ exist, then $x \vee y \vee z$ exists. A meet semilattice L is called a* median semilattice *if it is lower distributive and satisfies the median property.*

The concepts of semilattice, lattice, distributivity, and median are relevant to biocon-sensus. Consider the sets of binary relations ($\mathcal{R}_n$), equivalence relations ($\mathcal{E}_n$), weak orders ($\mathcal{O}_n$), hierarchies ($\mathcal{H}_n$), and weak hierarchies ($\mathcal{W}_n$), $\mathcal{X}_n$ being the set of all structures of the specified type that are defined on $S = S_n$. Table 5.1 gives the lattice-theoretic types of the posets $(\mathcal{X}_n, \subseteq)$; it is based on results such as the following lemma.

Lemma 5.7. *For $n \geq 1$ the poset $\mathcal{H} = (\mathcal{H}_n, \subseteq)$ is a median semilattice.*

Proof. Clearly $\mathcal{H}$ is a meet semilattice. For distributivity let $H, H_1, H_2, H_3 \in \mathcal{H}_n$ be such that $H_i \subseteq H$ for $i \in \{1, 2, 3\}$. For all $i, j \in \{1, 2, 3\}$, $H_i \cap H_j$ and $H_i \cup H_j$ are hierarchies

Table 5.1. *Ordered Sets in Bioconsensus*

Poset	Size	Type	Figure
$(\mathcal{R}_n, \subseteq)$	$n \geq 1$	Distributive lattice	5.3
$(\mathcal{E}_n, \subseteq)$	$n \geq 3$	Nondistributive lattice	5.4
$(\mathcal{O}_n, \subseteq)$	$n \geq 1$	Join semilattice	5.5 on page 82
$(\mathcal{H}_n, \subseteq)$	$n \geq 1$	Median semilattice	5.6 on page 83
$(\mathcal{W}_n, \subseteq)$	$n \geq 3$	Lower distributive non-median meet semilattice	5.7 on page 84

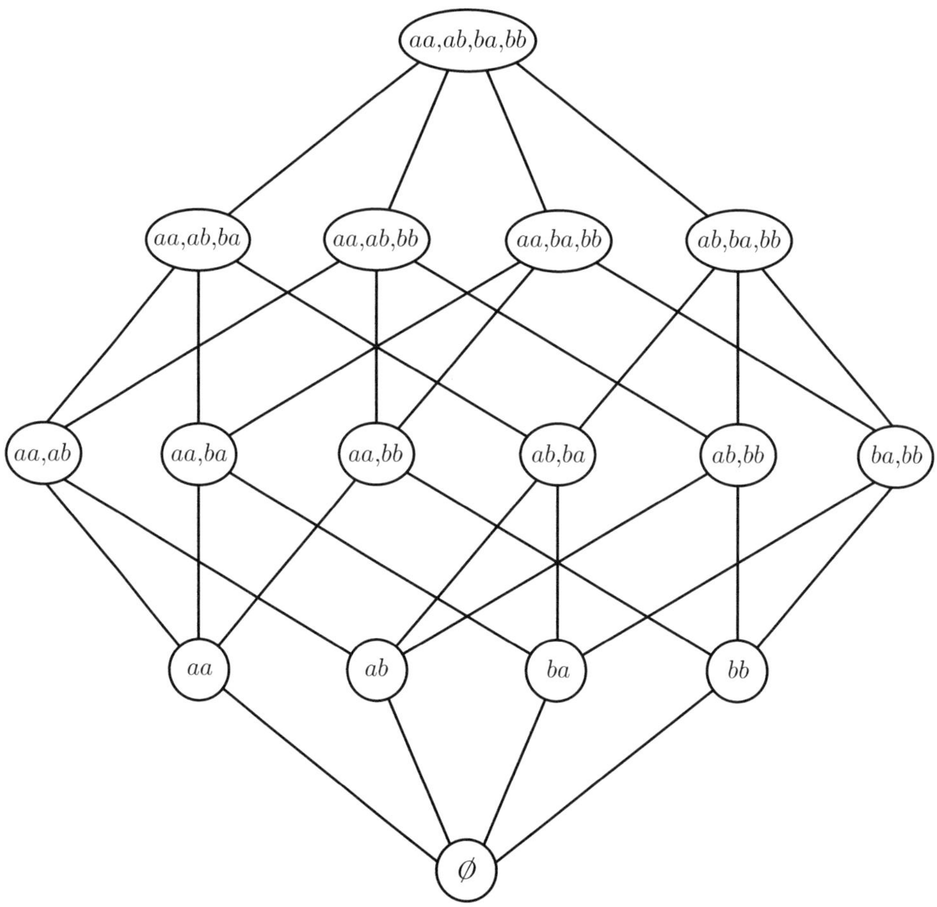

Figure 5.3. *Lattice* $(\mathcal{R}_2, \subseteq)$ *of Binary Relations on* $S = \{a, b\}$. *By the numbering of relations in Table* 1.4 *on page* 6, *the unit is the relation* $R_{15} = \{aa, ab, ba, bb\}$; *the zero is the relation* $R_0 = \emptyset$.

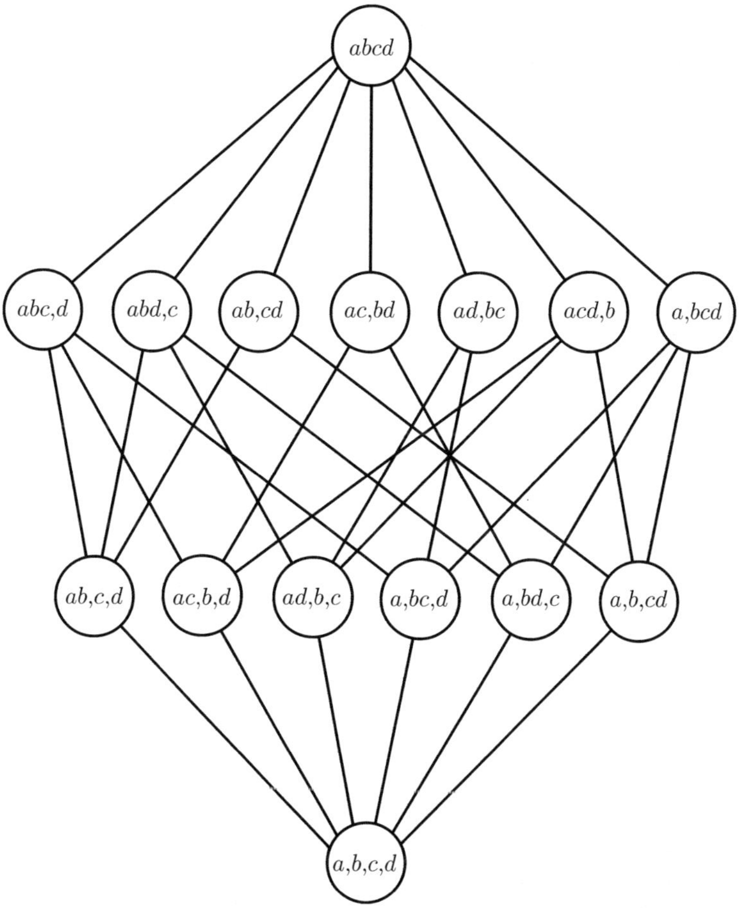

Figure 5.4. *Lattice $(\mathcal{E}_4, \subseteq)$ of Equivalence Relations on $S = abcd$. The leftmost node, which is labeled "abc, d," is the equivalence relation with abc and d as its equivalence classes.*

in $(H]$ and thus in $\mathcal{H}_n$. For all $\emptyset \neq X \subseteq S$,

$$\begin{aligned} X \in (H_1 \cup H_2) \cap H_3 &\iff X \in (H_1 \cup H_2) \wedge X \in H_3 \\ &\iff (X \in H_1 \vee X \in H_2) \wedge X \in H_3 \\ &\iff (X \in H_1 \wedge X \in H_3) \vee (X \in H_2 \wedge X \in H_3) \\ &\iff (X \in H_1 \cap H_3) \vee (X \in H_2 \cap H_3) \\ &\iff X \in (H_1 \cap H_3) \cup (H_2 \cap H_3), \end{aligned}$$

whence $(H]$ is distributive and $\mathcal{H}$ is lower distributive. For the median property let $H_1, H_2, H_3 \in \mathcal{H}_n$ be such that $H_{ij} = H_i \cup H_j \in \mathcal{H}_n$ for $i, j \in \{1, 2, 3\}$ with $i < j$, and set $H = H_1 \cup H_2 \cup H_3$. If $X, Y \in H$, then each separately is in at least two of the H_{ij},

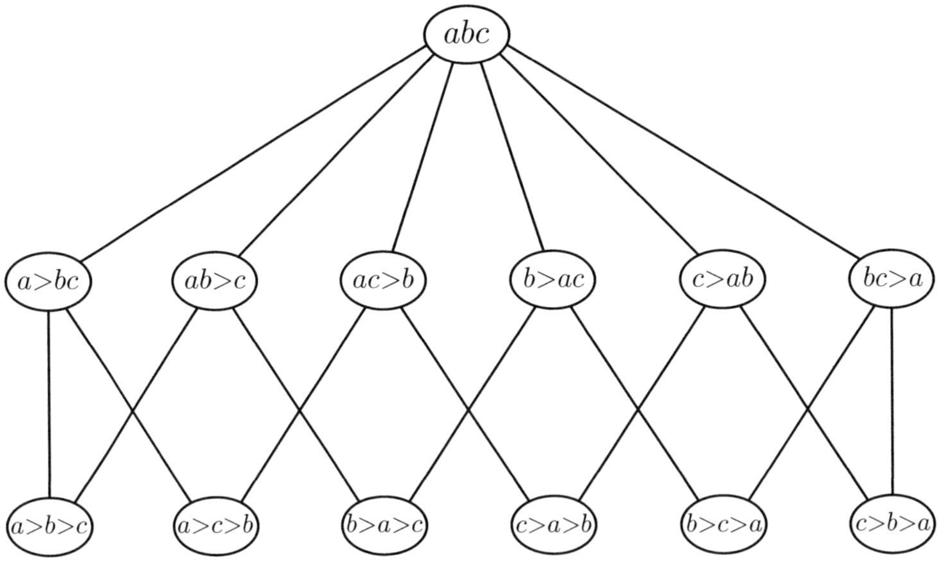

Figure 5.5. *Join Semilattice* $(\mathcal{O}_3, \subseteq)$ *of Weak Orders on* $S = abc$*. The middle left node, labeled "*$a > bc$*," is the weak order* $\{aa, ab, ac, bb, bc, cb, cc\}$.

so they together are in at least one of the H_{ij}; thus $X \cap Y = \{\emptyset, X, Y\}$, so that $H \in \mathcal{H}$, whence $\mathcal{H}$ satisfies the median property. □

The median property may be less familiar than distributivity.

Example 5.8. For $S = abcd$ let $H_1, H_2, H_3 \in \mathcal{H}_4$ with the nontrivial clusters in $H_1^* = \{ab\}$, $H_2^* = \{cd\}$, and $H_3^* = \{bcd\}$. If $H = H_1 \cup H_2 \cup H_3$, then $H^* = \{ab, cd, bcd\}$, so $H \notin \mathcal{H}_4$. It follows by the median property that $\{H_1 \cup H_2, H_1 \cup H_3, H_2 \cup H_3\} \not\subseteq \mathcal{H}_4$; indeed $(H_1 \cup H_3)^* = \{ab, bcd\}$, so that $H_1 \cup H_3 \notin \mathcal{H}_4$. ■

To characterize consensus rules for meet semilattices, we will exploit the structure of a set of lattice elements that are the irreducibles in terms of which all nontrivial elements can be described.

Definition 5.9. *An element* j *in a finite meet semilattice* L *is called* join irreducible *if* j *covers a unique element* $p(j) \in L$*, i.e.,* $p(j) < j$ *and* $(\forall x \in L)(p(j) \leq x < j \Longrightarrow p(j) = x)$*.* $J = J_L$ *is the set of join irreducibles of* L*, where always* $0 \notin J$*. Any* $j \in J$ *is called an* atom *if* j *covers* 0.

Example 5.10. In Figure 5.2 on page 79, $J_{L_1} = \{a, b, c\}$, c being an atom; $J_{L_2} = \{a, c, d, e\}$, d and e being atoms, but b is not join irreducible since $b = c \vee e = d \vee e$. In Figures 5.3, 5.4, 5.6, and 5.7, all of the join irreducibles are atoms. ■

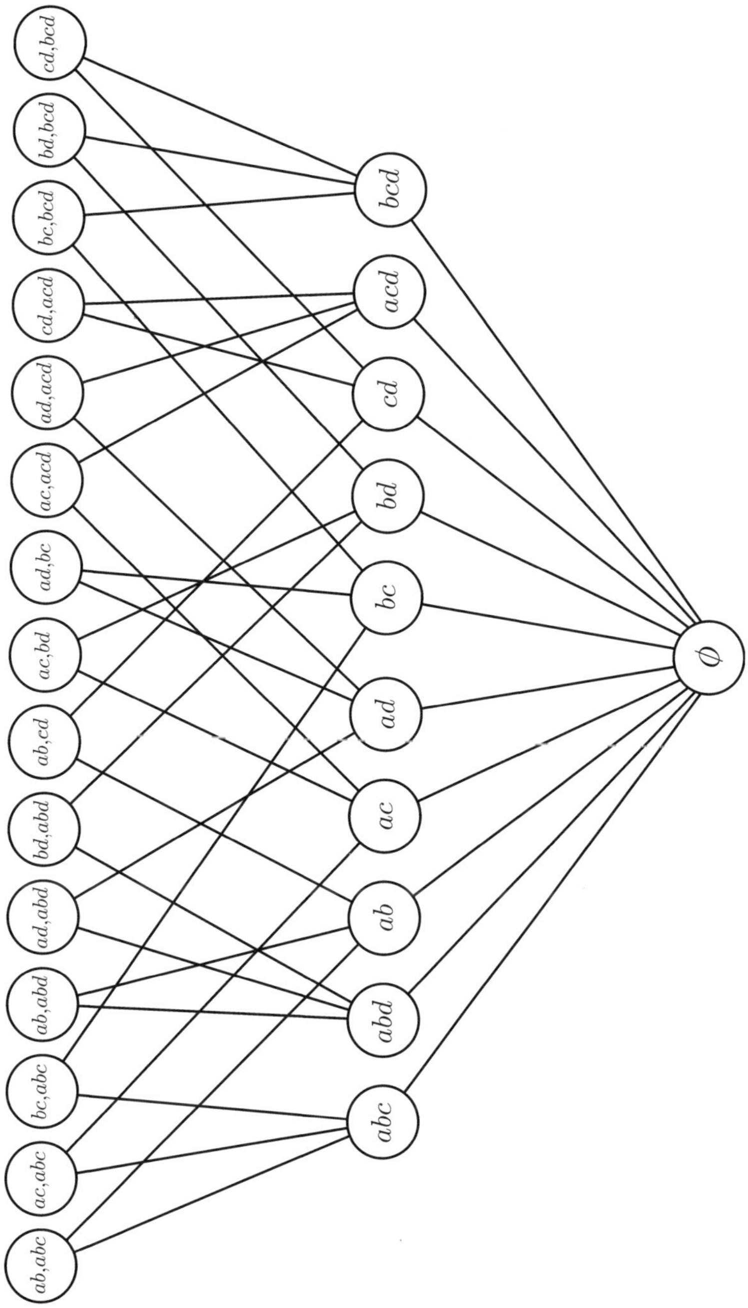

Figure 5.6. *Median Semilattice ($\mathcal{H}_4$, $\subseteq$) of Hierarchies on $S = abcd$. The top left node is the hierarchy H with $H^* = \{ab, abc\}$; the zero is the null hierarchy.*

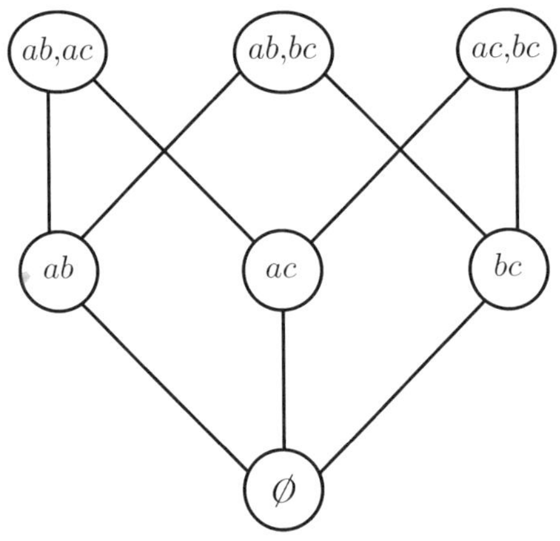

Figure 5.7. *Meet Semilattice $(\mathcal{W}_3, \subseteq)$ of Weak Hierarchies on $S = abc$. The top left node is the weak hierarchy W with $W^* = \{ab, ac\}$; the zero is the null hierarchy.*

Each join irreducible of $\mathcal{H}$ is a hierarchy H_X, where $\emptyset \subset X \subset S$ and X is the only nontrivial cluster of H_X, i.e., the join irreducibles can be identified with the nontrivial subsets of S. Every join irreducible of $\mathcal{H}$ covers $H_\emptyset$ and so is an atom. Analogous statements hold for $\mathcal{W}$. It may not be surprising that the properties of a meet semilattice L should determine the properties of the set J_L of join irreducibles, but which properties and how? In this context a dependence relation [300] between join-irreducible elements helps to identify several relevant structural properties.

Definition 5.11. Monjardet [296, pp. 55, 61]. *For each meet semilattice $L = (L, \leq)$ with set J of join irreducibles, let the relation $\delta \subseteq J^2$ be such that*

$$(\forall j, j' \in J)(j\delta j' \iff j \neq j' \wedge (\exists x \in L)((j, j' \not\leq x) \wedge (j < j' \vee x))).$$

J is called δ strongly connected *if for all $j, j' \in J$ a sequence $j = j_0, \ldots, j_p = j'$ of join irreducibles exists such that $j_{i-1}\delta j_i$ for all $i = 1, \ldots, p$.*

Example 5.12. For L_1 (Figure 5.2 on page 79) with $J_{L_1} = \{a, b, c\}$, then $c \not\leq 0$, $a \not\leq 0$, and $c < a \vee 0$, so that $c\delta a$, while $c \not\leq 0$, $b \not\leq 0$, and $c < b \vee 0$, so that $c\delta b$; indeed $\delta = \{ca, cb\}$. For L_2 (Figure 5.2) with $J_{L_2} = \{a, c, d, e\}$, then $\delta = \{ca, cd, ce, da, dc\}$. For $\mathcal{R}_2$ (Figure 5.3 on page 80) and $\mathcal{H}_4$ (Figure 5.6 on page 83), $\delta = \emptyset$. For $\mathcal{E}_4$ (Figure 5.4 on page 81), every pair of distinct join irreducibles is in δ. These features hold for other $\mathcal{R}_n$, $\mathcal{H}_n$, and $\mathcal{E}_n$. ■

Lemma 5.13. *Let $(\mathcal{X}_n, \subseteq)$ have J as its set of join irreducibles.*

(i) *If $n \geq 2$ for $(\mathcal{R}_n, \subseteq)$, then $\delta = \emptyset$.*
(ii) *If $n \geq 3$ for $(\mathcal{H}_n, \subseteq)$, then $\delta = \emptyset$.*
(iii) *If $n \geq 3$ for $(\mathcal{E}_n, \subseteq)$, then $\delta = \{jj' \in J^2 : j \neq j'\}$ and J is δ strongly connected.*

Proof. *Concerning* (i). Consider any $j, j' \in J$ with $j \neq j'$. The elements of S can be relabeled so that $j = \{ab\}$ and $j' = \{cd\}$ with $a \neq c$ or $b \neq d$. Since $n \geq 2$ choose any $x \in \mathcal{R}_n$ such that $ab \notin x$ and $cd \notin x$; then $j = \{ab\} \nsubseteq \{cd\} \cup x = j' \vee x$, so that $jj' \notin \delta$, from which the result follows.

Concerning (ii). A similar argument applies.

Concerning (iii). Consider any $j, j' \in J$ with $j \neq j'$. The elements of S can be relabeled so that either: $ab \in j$ and $ac \in j'$, whence choose $x \in J$ with $bc \in x$; or $ab \in j$ and $cd \in j'$, whence choose $x \in \mathcal{E}_n$ with $ab \notin x$, $cd \notin x$, and $\{ad, bc\} \subseteq x$. Either way, $j < j' \vee x$, so that $jj' \in \delta$, from which the result follows. □

The structure of δ determines, and is determined by, the distributivity of L.

Proposition 5.14. Monjardet [296, pp. 54–55]. *If $L = (L, \leq)$ is a meet semilattice, then these are equivalent:*

(i) *L is lower distributive.*
(ii) $(\forall j, j' \in J_L)(\forall x \in L)((j \neq j') \wedge (j < j' \vee x) \Longrightarrow (j < j') \vee (j \leq x))$.
(iii) $(\forall j, j' \in J_L)(j\delta j' \Longleftrightarrow j < j')$.

From Lemma 5.13 and Proposition 5.14 we have that $\mathcal{R}_n$ is distributive for $n \geq 2$, $\mathcal{H}_n$ is lower distributive for $n \geq 3$, but $\mathcal{E}_n$ is not distributive for $n \geq 3$.

5.2 Semilattice Rules

Let $C : L^k \longrightarrow L$ be a consensus rule on a meet semilattice $L = (L, \leq)$. We simplify notation when no confusion arises.

Definition 5.15. *Let L be a meet semilattice with J as the set of join-irreducibles, 0 as the zero, and 1 as the unit when it exists. For all $j \in J$, $x \in L$, and $P = (x_1, \ldots, x_k) \in L^k$,*

shorten	*to*
$C(P)$	CP
$(x, \ldots, x) \in L^k$	$(x)_k$
$\{i \in K : j \leq x_i\}$	$K_j(P)$.

Consider what axioms would be suitable to describe such consensus rules on L. The formulations in Table 5.2 of axioms for meet semilattices include plausible translations of many of the axioms in Chapters 2, 3, and 4. Concepts of autonomy, independence, monotonicity, neutrality, symmetry, and unanimity are all represented; but where before we asked if a set X were an element of a set H of sets, now we ask if a join irreducible j is a lower bound of a semilattice element x. On such meet semilattices are defined two natural families of consensus rules, which are based on concepts of oligarchy and decisiveness.

Table 5.2. *Axioms: Rules on Meet Semilattices* I. *For notation see Definition* 5.15.

***Atn*: Autonomy** $(\forall j \in J)(\exists P \in L^k)(j \leq CP)$

***Cst$_0$*: 0-Constant** $(\forall P \in L^k)(CP = 0)$

***Cst$_1$*: 1-Constant** $(\forall P \in L^k)(CP = 1)$

***DM*: Decisive Monotonicity**
$(\forall j \in J)(\forall P, P' \in L^k)(K_j(P) \subseteq K_j(P') \Longrightarrow [j \leq CP \Longrightarrow j \leq CP'])$

***DN*: Decisive Neutrality**
$(\forall j, j' \in J)(\forall P, P' \in L^k)(K_j(P) = K_{j'}(P') \Longrightarrow [j \leq CP \Longleftrightarrow j' \leq CP'])$

***Ext*: Extensiveness** $(\forall x \in L)(x \leq C((x)_k))$

***Idm*: Idempotence** $(\forall x \in L)(x = C((x)_k))$

***Idm$_1$*: 1-Idempotence** $C((1)_k) = 1$

***Idm$_b$*: Bi-idempotence**
$(\forall x, y \in L)(\forall P \in L^k)(\{x_i : i \in K\} \subseteq \{x, y\} \Longrightarrow CP \in \{x, y\})$

***Ind*: Independence**
$(\forall j \in J)(\forall P, P' \in L^k)(K_j(P) = K_j(P') \Longrightarrow [j \leq CP \Longleftrightarrow j \leq CP'])$

***Ist*: Isotony** $(\forall P, P' \in L^k)((\forall i \in K)(x_i \leq x_i') \Longrightarrow CP \leq CP')$

***MC*: Meet Compatibility** $(\forall P \in L^k)(x_1 \wedge \cdots \wedge x_k \leq CP)$

***MN*: Monotonic Neutrality**
$(\forall j, j' \in J)(\forall P, P' \in L^k)(K_j(P) \subseteq K_{j'}(P') \Longrightarrow [j \leq CP \Longrightarrow j' \leq CP'])$

***Ntr*: Neutrality**
$(\forall j, j' \in J)(\forall P \in L^k)(K_j(P) = K_{j'}(P) \Longrightarrow [j \leq CP \Longleftrightarrow j' \leq CP])$

***Sym*: Symmetry** $(\forall P \in L^k)(\forall \text{ permutations } \sigma \text{ of } K)(CP = C(x_{\sigma(1)}, \ldots, x_{\sigma(k)}))$

***Unn*: Unanimity** $(\forall j \in J)(\forall P \in L^k)(K_j(P) = K \Longrightarrow j \leq CP)$

5.2.1 Meet Projection Rules

Definition 5.16. [296, p. 56]. *For all $A \subseteq K$, the A-*meet projection *consensus rule on the meet semilattice L is a function $C_A : L^k \longrightarrow L$ such that*

$$(\forall P = (x_1, \ldots, x_k) \in L^k)\left(C_A P = \bigwedge_{i \in A} x_i\right).$$

Meet projection rules are the lattice-theoretic analogues of the oligarchic rules on equivalence relations (Definition 3.7 on page 29). In the case where the set of join irreducibles is δ strongly connected (Definition 5.11 on page 84), these rules can be characterized.

Theorem 5.17. Monjardet [296, p. 66]. *Let L be a meet semilattice, J its set of join-irreducibles, J being δ strongly connected, and $C : L^k \longrightarrow L$ a consensus rule. Consider the following conditions:*

1. *C is an A-meet projection consensus rule with $\emptyset \subset A \subseteq K$.*
2. *C satisfies MN and $\neg Cst_0$.*
3. *C satisfies Ind and either Unn or MC or Ext.*
4. *C satisfies Ind, Ntr, Ist, and $\neg Cst_0$.*
5. *C satisfies Ind and Idm_1.*
6. *C is a meet projection consensus rule.*

If L is a meet semilattice that is not a lattice, then conditions 1–3 *are equivalent. If L is a lattice, then conditions* 2–6 *are equivalent. Moreover $|A| = k$ if and only if C satisfies Sym, and $|A| = 1$ if and only if C satisfies Idm_b.*

Because of Lemma 5.13(iii), Theorem 5.17 yields a characterization of meet projection consensus rules for equivalence relations.

Corollary 5.18. *Let C be a consensus rule on $L = (\mathcal{E}_n, \subseteq)$ with $n \geq 3$ and J its set of join irreducibles. C satisfies Ind and Idm_1 if and only if C is a meet projection consensus rule.*

Since meet projection rules are the lattice-theoretic equivalent of oligarchic consensus rules, and since 1-idempotence is a weak form of Pareto optimality, Corollary 5.18 is analogous to Mirkin's characterization of oligarchic rules on equivalence relations (Theorem 3.8 on page 30).

5.2.2 Federation Rules

The approach in section 4.1.1, where consensus rules on hierarchies were based on decisive families (Definition 4.2 on page 55), also can be used in a lattice-theoretic context.

Definition 5.19. *A set $\mathcal{F} \subseteq 2^K$ is called a* federation *on K if*

$$(\forall I \in \mathcal{F})(\forall J \subseteq K)(I \subseteq J \Longrightarrow J \in \mathcal{F}).$$

A set $\mathcal{F} \subseteq 2^K$ is called transversal *if*

$$(\forall I, J \in \mathcal{F})(I \cap J \neq \emptyset).$$

Thus the concepts of transversal federation and decisive family are equivalent.

Definition 5.20. [296, pp. 56–57]. *For each federation $\mathcal{F}$ on K, a* federation *consensus rule on L is a partial function $C_{\mathcal{F}} : L^k \longrightarrow L$ such that*

$$(\forall P = (x_1, \ldots, x_k) \in L^k)\left(C_{\mathcal{F}}P = \bigvee_{V \in \mathcal{F}} \bigwedge_{i \in V} x_i\right).$$

Federation rules are the lattice-theoretic analogues of the decisive-family consensus rules on hierarchies (Definition 4.6 on page 56). If L is a lower distributive meet semilattice, then $C_{\mathcal{F}}$ is well defined for each transversal federation $\mathcal{F}$, and if L is a lattice, then $C_{\mathcal{F}}$ is well defined for each federation $\mathcal{F}$. In these cases federation consensus rules can be characterized.

Theorem 5.21. Monjardet [296, p. 67]. *Let L be a lower distributive meet semilattice, J its set of join irreducibles, and $C : L^k \longrightarrow L$ a consensus rule. Consider the following conditions:*

1. *C is a federation consensus rule $C_{\mathcal{F}}$ with $\mathcal{F}$ a transversal federation.*
2. *C satisfies MN.*
3. *C satisfies DM and DN.*
4. *C satisfies Ind, Ntr, and Ist.*
5. *C is a federation consensus rule $C_{\mathcal{F}}$.*

If L is not a lattice, then conditions 1–3 *are equivalent. If L is a (distributive) lattice, then conditions* 2–5 *are equivalent.*

This yields characterizations of decisive-family consensus rules on hierarchies.

Corollary 5.22. *Let C be an idempotent consensus rule on $L = (\mathcal{H}_n, \subseteq)$ with $n \geq 3$ and J its set of join irreducibles. These are equivalent:*

(i) *C satisfies MN.*
(ii) *C satisfies DM and DN.*
(iii) *C is a federation consensus rule $C_{\mathcal{F}}$ with $\mathcal{F}$ a decisive family.*

Since requiring idempotence in Corollary 5.22 ensures that C is not constant, autonomy (*Atn*) is a consequence of monotonic neutrality (*MN*) [296, pp. 59–60], so Corollary 5.22 is analogous to McMorris and Neumann's characterization of decisive family rules on hierarchies (Theorem 4.9 on page 56).

5.3 Median Rules

How might the concrete idea of median rule on hierarchies (Definition 4.44 on page 70) be extended to an abstract order-theoretic setting? Since $\mathcal{H} = (\mathcal{H}_n, \subseteq)$ is a median semilattice (Lemma 5.7 on page 79), and since meet semilattices support meaningful measures of distance (Monjardet [294]), all that remains is to notice that the method of median rule for hierarchies adumbrates the general concept of a median consensus rule for median semilattices.

Definition 5.23. *Let* $L = (V, \leq)$ *be a finite median semilattice. The* distance $d(u, v)$ *between any* $u, v \in V$ *is the length of a geodesic, or shortest path, between* u *and* v *in the covering graph of* L*. A* median *of* $P = (v_1, \ldots, v_k) \in V^k$ *is an element* $v \in V$ *for which the* remoteness $r(v, P) = \sum_{i=1}^{k} d(v, v_i)$ *is minimum on* V*. The method of* median rule *for* $L = (V, \leq)$ *is a complete multiconsensus rule* $Med : V^* \longrightarrow 2^V \setminus \{\emptyset\}$ *such that*

$$(\forall P \in V^*)(Med(P) = Med\, P = \{v \in V : v \text{ is a median of } P\}).$$

The median rule on a median semilattice has close ties with graph theory since the underlying distance measure is defined on the semilattice's covering graph.

Definition 5.24. *Let* $G = (V, E)$ *be a finite connected graph. The* distance $d(u, v)$ *between any* $u, v \in V$ *is the length of a geodesic, or shortest path, between* u *and* v *in* G*. A* median *of* $P = (v_1, \ldots, v_k) \in V^k$ *is a vertex* $v \in V$ *for which the* remoteness $r(v, P) = \sum_{i=1}^{k} d(v, v_i)$ *is minimum on* V*. The method of* median rule *for* $G = (V, E)$ *is a complete multiconsensus rule* $Med : V^* \longrightarrow 2^V \setminus \{\emptyset\}$ *such that*

$$(\forall P \in V^*)(Med(P) = Med\, P = \{v \in V : v \text{ is a median of } P\}).$$

If $G = (V, E)$ and $P \in V^*$ are not further constrained, then medians in G need not be unique: in the covering graph of Figure 5.1 on page 78, the profile $P = (c, d)$ has $Med\, P = V$. Consider the prospects for unique medians in arbitrary graphs when profiles are short. For all $P = (v) \in V^1$, then $Med\, P = \{v\}$, so medians are trivially unique. Let the *interval* in G between any $u, v \in V$ be the set

$$I(u, v) = \{w \in V : d(u, v) = d(u, w) + d(w, v)\};$$

for all $P = (u, v) \in V^2$, then $Med\, P = I(u, v)$, so medians are unique if and only if $u = v$. Although medians need not exist for arbitrary $P = (u, v, w) \in V^3$ in arbitrary graphs, yet graphs exist in which every such profile has a unique median.

Definition 5.25. *A connected graph* $G = (V, E)$ *is called a* median graph *if* $|Med\, P| = 1$ *for all* $P \in V^3$*. Equivalently, a connected graph* $G = (V, E)$ *is a median graph if* $|I(u, v) \cap I(u, w) \cap I(v, w)| = 1$ *for all* $u, v, w \in V$*.*

Example 5.26. These are median graphs (Figure 5.8): (1) Any tree, in which every two vertices are joined by a unique path. (2) Any covering graph of a finite distributive lattice. (3) Any *n-cube*, Q_n, in which $\{0, 1\}^n$ is the vertex set and two vertices are adjacent if they differ in exactly one place. For all $u = u_1 u_2 \cdots u_n$, $v = v_1 v_2 \cdots v_n$, $w = w_1 w_2 \cdots w_n \in \{0, 1\}^n$, the median $x = x_1 x_2 \cdots x_n$ of (u, v, w) is determined by the majority rule: for all $i \in N$, $x_i = \epsilon$ if ϵ occurs at least twice among u_i, v_i, w_i. ■

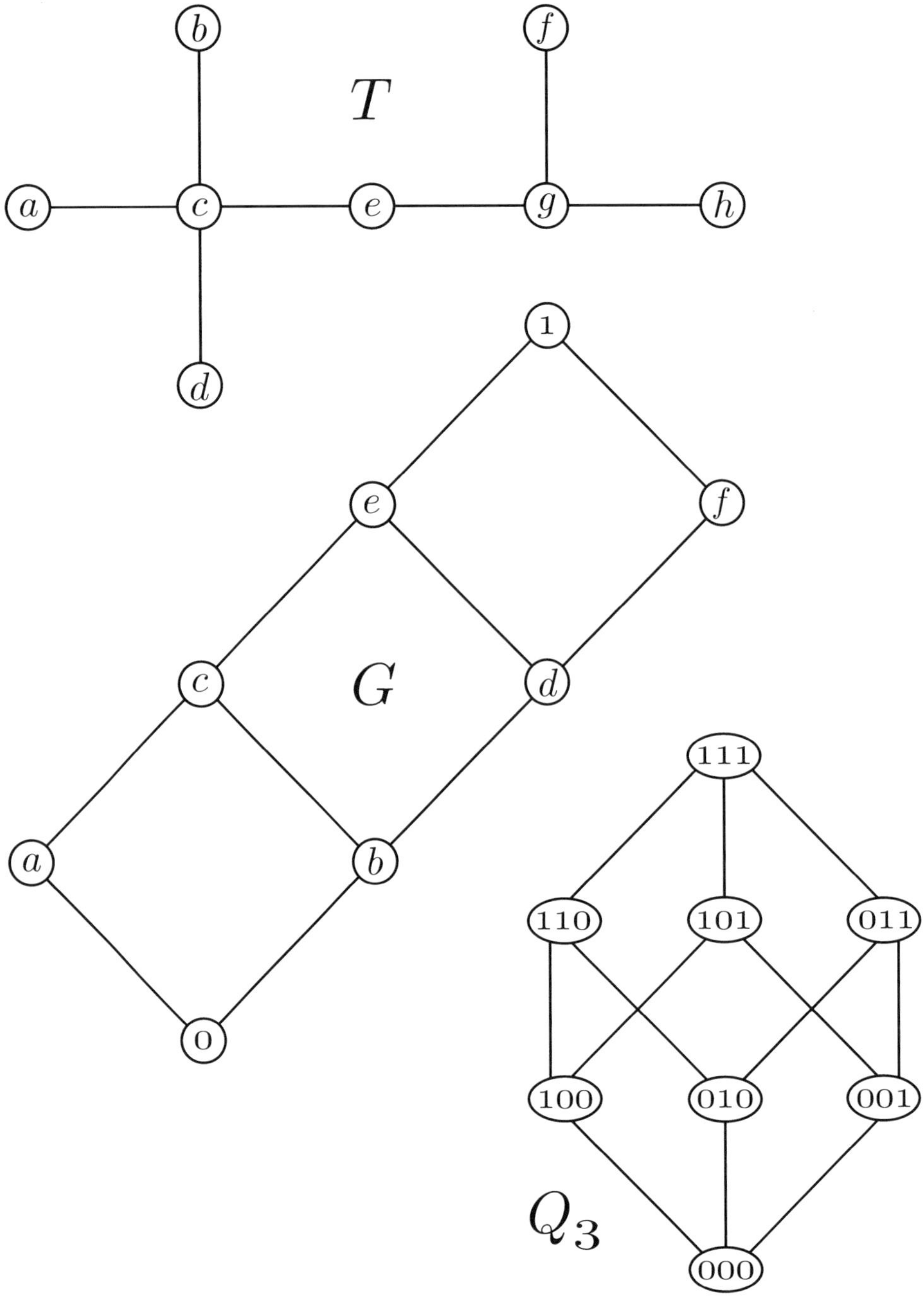

Figure 5.8. *Median Graphs: a tree T, a covering graph G of a distributive lattice, an n-cube Q_3. Example 5.26 explains.*

Table 5.3. *Splits $\Psi_i = (G_1, G_2)$ of the median graph $G = (V, E)$ in Figure* 5.8.

Split	Vertices in G_1	Vertices in G_2	F_{12}
Ψ_1	0bdf	1ace	{0a, bc, de, 1f}
Ψ_2	0a	1bcdef	{0b, ac}
Ψ_3	0abc	1def	{bd, ce}
Ψ_4	0abcde	1f	{df, 1e}

Median graphs were introduced independently by Avann [22], Nebeský [312], and Mulder and Schrijver [306]. They are relevant to problems of location theory [196, 281, 147] since for all vertices u, v, w of such a graph, the median x is on a geodesic between each pair of the three: x might be a good site from which to service facilities located at u, v, and w. In section 5.3.1 we characterize the median rule on median graphs, and in section 5.3.2 we derive from that result a useful characterization of the median rule on median semilattices.

5.3.1 Median Graphs

An effective way to analyze median rules on median graphs involves decomposing the median graph into subgraph pairs called splits.

Definition 5.27. *Let $G = (V, E)$ be a median graph. For all $v_1v_2 \in E$ let G_1 be the subgraph of G induced by all vertices nearer to v_1 than to v_2, and let G_2 be the subgraph of G induced by all vertices nearer to v_2 than to v_1. Then $\Psi = (G_1, G_2)$ partitions G and is called a* split*; F_{12} is the set of edges between G_1 and G_2. For $i \in \{1, 2\}$ let G_{0i} be the subgraph induced by the endvertices in G_i of the edges in F_{12}. If $P \in V^*$, for $i \in \{1, 2\}$ let P_i be the subprofile of P having (in the same order) all those vertices of P lying in G_i. For P the split $\Psi = (G_1, G_2)$ is called* equal *if $|P_1| = |P_2|$; otherwise it is* unequal. *When no confusion will arise, G_i is also used to denote the graph's set of vertices.*

Example 5.28. Table 5.3 describes the splits of the median graph G in Figure 5.8. For $P = (b, a, d, a)$, splits Ψ_1 and Ψ_2 are equal but Ψ_3 and Ψ_4 are unequal. ■

The split concept lets us explain the finer structure of the median rule on median graphs. Recall for two graphs $G_1 = (V_1, E_1)$ and $G_2 = (V_2, E_2)$ that the *union* $G_1 \cup G_2$ is the graph with vertex set $V_1 \cup V_2$ and edge set $E_1 \cup E_2$, while the *intersection* $G_1 \cap G_2$ is the graph with vertex set $V_1 \cap V_2$ and edge set $E_1 \cap E_2$. A graph G is called *cube-free* if the cube Q_3 (Figure 5.8) does not occur as an induced subgraph of G.

Proposition 5.29. *Let $G = (V, E)$ be a median graph.*

(i) [264, p. 224] *Let $P \in V^*$ and let (G_1, G_2) be an equal split of G.*
$(\forall uv \in F_{12})(u \in Med P \iff v \in Med P)$.

(ii) [264, p. 225] *Let (G_1, G_2) be a split of G and let v be a vertex of G_2. There exists a split (H_1, H_2) with v in H_{02}, $G_1 \subseteq H_1$, and $H_2 \subseteq G_2$.*

Table 5.4. *Axioms: Rules on Median Graphs. For notation see Definitions* 5.27 *and* 5.30.

***Btw*: Betweenness** $(\forall P = (v_1, v_2) \in V^2)(CP = I(v_1, v_2))$

***Cnd*: Condorcet** $(\forall P \in V^*)(\forall$ equal splits (G_1, G_2) of G for $P)(\forall v_1 v_2 \in F_{12})$
$(v_1 \in CP \Longleftrightarrow v_2 \in CP)$

***Css*: Consistency** $(\forall P, P' \in V^*)(CP \cap CP' \neq \emptyset \Longrightarrow C(PP') = CP \cap CP')$

***Fth*: Faithfulness** $(\forall P = (v_1) \in V^1)(CP = \{v_1\})$

***PI*: Population Invariance** $(\forall P, P' \in V^*)(\{P\} = \{P'\} \Longrightarrow CP = CP')$

***QC*: Quasi-Consistency** $(\forall P, P' \in V^*)(CP = CP' \Longrightarrow C(PP') = CP)$

***Sym*: Symmetry** $(\forall P \in V^k)(\forall$ permutations σ of $K)(CP = C(v_{\sigma(1)}, \dots, v_{\sigma(k)}))$

(iii) [264, p. 225] $\forall P \in V^* : Med\, P = \bigcap\{G_1 : (G_1, G_2)$ *is a split with* $|P_1| > |P_2|\}$.

(iv) [266, p. 172] *Let* $P \in V^*$ *be of even length such that every split of* G *is equal for* P. *Then* $Med\, P = V$.

(v) [266, p. 173] *For all* $P = (v_1, \dots, v_k) \in V^k$ *let* $P - v_i$ *be the profile* $P = (v_1, \dots, v_{i-1}, v_{i+1}, \dots, v_k) \in V^{k-1}$ *in which* v_i *has been deleted. If* $k = 2m+1 > 1$, *then* $Med\, P = \cap_{i=1}^{k} Med(P - v_i)$.

(vi) [266, p. 177] *If* G *is cube-free and* $P \in V^k$ *for* $k = 2m > 1$, *then there exists a permutation* σ *of* $\{1, \dots, 2m\}$ *with* $\sigma P = (v_1, \dots, v_{2m})$ *such that* $Med\, P = \cap_{i=1}^{m} I(v_{2i-1}, v_{2i})$.

Let $C : V^* \longrightarrow 2^V \setminus \{\emptyset\}$ be a consensus rule on a median graph $G = (V, E)$. We simplify notation when no confusion arises.

Definition 5.30. *For all* $P, P' \in V^*$,

shorten	*to*
$C(P)$	CP
concatenation of P *and* P'	PP'
set of vertices in P	$\{P\}$.

Consider what axioms would be suitable to describe such consensus rules on G. The formulations in Table 5.4 are graph-theoretic analogues of corresponding axioms for hierarchies in Table 4.2 on page 62 and Table 4.3 on page 73. The axioms vary in degree of generality: although consistency, faithfulness, and symmetry make sense for consensus rules on any graph, betweenness requires that its graph be connected, while the condorcet axiom requires furthermore that its graph support the concept of split. Excepting the condorcet axiom these axioms embody simple concepts, yet in combination they yield interesting and nontrivial results.

Theorem 5.31. McMorris, Mulder, and Roberts [266, p. 178]. *Let $C : V^* \longrightarrow 2^V \setminus \{\emptyset\}$ be a consensus rule on a cube-free median graph $G = (V, E)$. $C = Med$ if and only if C satisfies Btw, Css, and Sym.*

Proof. It is easy to show that *Med* satisfies *Btw*, *Css*, and *Sym* on such graphs, so instead let C satisfy the three axioms. We will use induction on the length k of P to show that $CP = Med\,P$ for all $P \in V^*$. For $k = 1$, *Btw* and *Css* show that $C((v)) = C((v)) \cap C((v)) = C((v, v)) = I(v, v) = \{v\} = Med((v))$. Let $k = 2m + 1 > 1$. Then $Med\,P = \cap_{i=1}^{k} Med(P - v_i)$ by Proposition 5.29(v), so that $Med\,P = \cap_{i=1}^{k} C(P - v_i)$ by induction. Since $Med\,P \neq \emptyset$, *Sym* and repeated use of *Css* yield $Med\,P = C((v_1, \ldots, v_1, v_2, \ldots, v_2, \ldots, v_k, \ldots, v_k))$, with each vertex v_i of P appearing exactly $2m$ times. Using *Sym* and *Css* again yields $CP = Med\,P$. Finally let $k = 2m > 1$. Since G is cube-free, use Proposition 5.29(vi) to write $P = (v_1, \ldots, v_{2m})$ in such a way that $Med\,P = \cap_{i=1}^{m} I(v_{2i-1}, v_{2i})$. Then $Med\,P = \cap_{i=1}^{m} C((v_{2i-1}, v_{2i}))$ by *Btw*, whence $CP = Med\,P$ by *Css* and *Sym*. □

Open Problem 5.32. Give an example of a consensus rule on a median graph that satisfies *Btw*, *Css*, and *Sym* but is not the median rule. ■

Open Problem 5.33. Do there exist other classes of graphs (or metric spaces) for which the median rule can be characterized by *Btw*, *Css*, and *Sym*? ■

To characterize median rules on median graphs in the general case, one can invoke the condorcet axiom in an argument using concepts of convexity in graphs.

Definition 5.34. *A set $W \subseteq V$ of graph $G = (V, E)$ is called* convex *if $I(u, v) \subseteq W$ for all $u, v \in W$. A subgraph of G is called* convex *if it is induced by a convex set of vertices of G, so any convex subgraph of a connected graph is connected, and any intersection of convex sets (or subgraphs) is convex.*

Theorem 5.35. McMorris, Mulder, and Powers [264, p. 226]. *Let $C : V^* \longrightarrow 2^V \setminus \{\emptyset\}$ be a consensus rule on a median graph $G = (V, E)$. $C = Med$ if and only if C satisfies Cnd, Css, and Fth.*

Proof. Let $G = (V, E)$ be a median graph. It is easy to see that *Med* satisfies *Css* [46, p. 310] and *Fth* on G; Proposition 5.29(i) shows that *Med* satisfies *Cnd* on G. For the converse let rule $C : V^* \longrightarrow 2^V \setminus \{\emptyset\}$ satisfy *Cnd*, *Css*, and *Fth* on G. First we will prove that if (G_1, G_2) is an unequal split with $|P_1| > |P_2|$, then $CP \subseteq G_1$. Let the contrary hold, so that $u \in G_2$ for some $u \in CP$. Because of Proposition 5.29(ii), let (H_1, H_2) be a split with u in H_{02}, $G_1 \subseteq H_1$, and $H_2 \subseteq G_2$, and let v be the neighbor of u in H_{01}; then $|P(H_1)| \geq |P_1| > |P_2| \geq |P(H_2)|$, so (H_1, H_2) is an unequal split for P. Let $|P| = k$ and $|P(H_1)| = p$, so that $p > k - p$, whence $2p - k > 0$. Then let $Q = P \cdot (u)_{2p-k}$ be the concatenation of P with the profile having $2p-k$ copies of u, whence $|Q(H_1)| = |Q(H_2)|$, so that (H_1, H_2) is an equal split for Q. Since $u \in CP$ and $C((u, \ldots, u)) = \{u\}$ by *Fth* and *Css*, then $CQ = \{u\}$ by *Css*; but *Cnd* implies that $u \in CQ \iff v \in CQ$, a contradiction, whence $CP \subseteq G_1$. It follows by Proposition 5.29(iii) that $CP \subseteq Med\,P$. Since $Med\,P$ is the intersection of convex subgraphs, it is convex and induces a connected subgraph. Let uv be any edge in $Med\,P$, with (H_1, H_2) the split associated with uv; then Proposition 5.29(iii)–(iv) imply that (H_1, H_2) is an equal split of G for P. Since C satisfies *Cnd*, then $u \in CP \iff v \in CP$; but since $\emptyset \subset CP \subseteq Med\,P$, the connectivity of $Med\,P$ implies that $CP = Med\,P$. □

5.3.2 Distributive Semilattices

To understand the relationships between median graphs and median semilattices, consider the ordering of a graph's vertices relative to a given vertex.

Definition 5.36. *For each finite median graph $G = (V, E)$ and fixed vertex $z \in V$, let the* canonical order $\leq_z$ *of G with respect to z be the partial order on V such that $(\forall u, v \in V)(u \leq_z v \iff u \in I(z, v))$.*

Example 5.37. The median graph $G = (V, E)$ in Figure 5.8 on page 90 is the covering graph of a distributive lattice $L = (V, \leq)$ that has $0 \in V$ as its zero; thus $\leq$ is identical to the canonical order $\leq_0$ of G with respect to 0. Figure 5.9 shows G as the covering graph of the canonical orders $\leq_a$ and $\leq_b$. ■

Canonical orders induce well-known correspondences between median graphs and median semilattices.

Proposition 5.38. [27, p. 133]. *The covering graph of any finite median semilattice is a median graph; conversely every finite median graph yields a median semilattice with respect to any canonical order $\leq_z$.*

Furthermore a gating concept reveals a subtle relationship between the splits of a median graph and the join irreducibles of any corresponding median semilattice.

Definition 5.39. *For each graph $G = (V, E)$ with $W \subseteq V$ and $u \in V$, a vertex $z \in W$ is called a* gate for u in W *if $z \in I(u, v)$ for all $v \in W$.*

Vertex $u \in V$ has at most one gate in any set $W \subseteq V$; if z is u's gate in W, then z is u's unique nearest vertex in W.

Example 5.40. For the median graph $G = (V, E)$ in Figure 5.8 on page 90 let $W = abd$. Then f has the gate d in W since

$$I(f, a) \cap I(f, b) \cap I(f, d) = V \cap bdf \cap df = df;$$

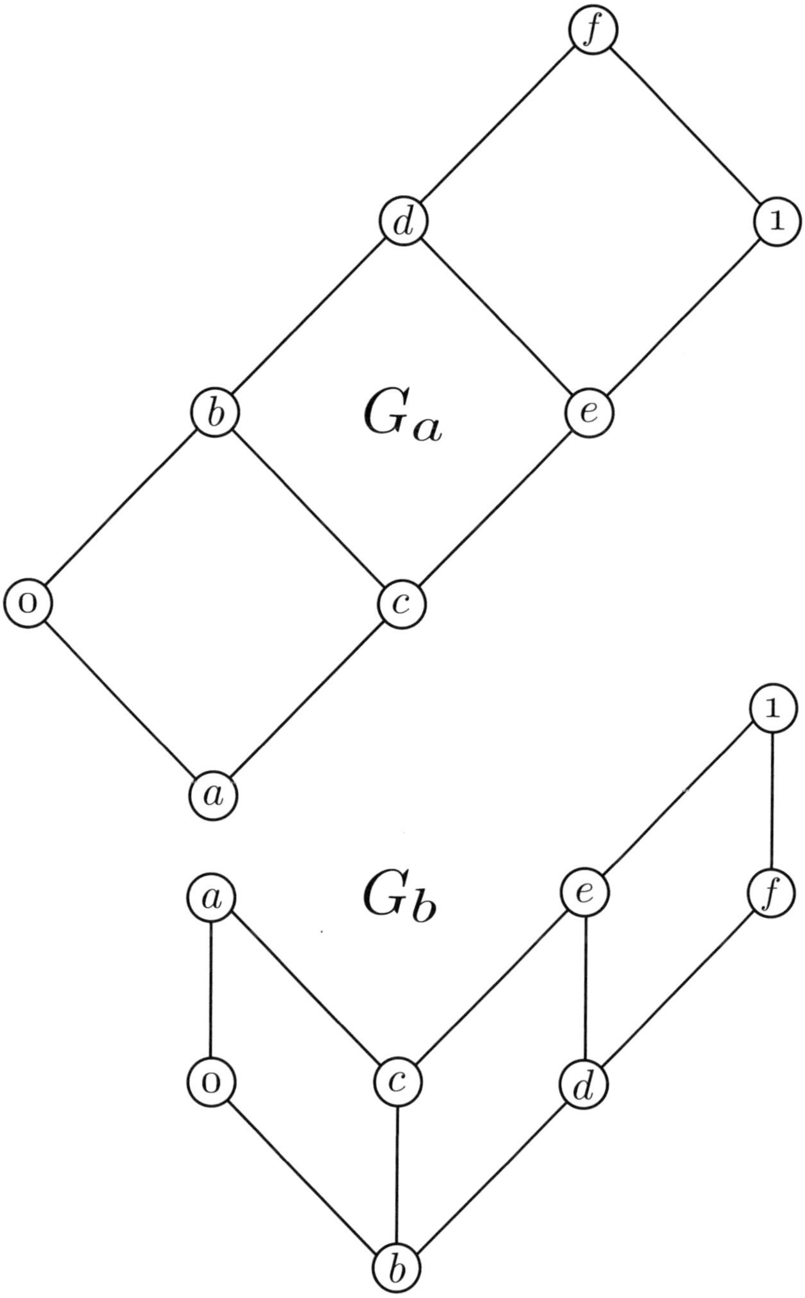

Figure 5.9. *Canonical Orders of a Median Graph. For the median graph G of Figure* 5.8 *on page* 90 *are shown the covering graphs of its canonical orders* $\leq_a$ *and* $\leq_b$.

but $e \in V$ has no gate in W since

$$I(e, a) \cap I(e, b) \cap I(e, d) = ace \cap bcde \cap de = e. \quad \blacksquare$$

If (G_1, G_2) is a split of a median graph, however, then each vertex of G_1 (respectively, G_2) has a gate in G_2 (respectively, G_1), and consequently we have the following proposition.

Proposition 5.41. [264, p. 227] *Let $G = (V, E)$ be a median graph.*

(i) *Let (G_1, G_2) be a split of G, u a vertex of G_1, and z the gate of u in G_2. Then the neighbor w_z of z in G_1 is the gate of u in G_{01} as well as in $G_{01} \cup G_2$.*
(ii) *Let $u \in V$. For each split (G_1, G_2) of G with $u \in G_1$, the gate z of u in G_2 is the unique join irreducible in G_{02} in the median semilattice $(V, \leq_u)$.*

Example 5.42. The median graph $G = (V, E)$ in Figure 5.8 on page 90 has the four splits in Example 5.28 on page 91.

(i) Consider the split $\Psi_3 = (G_1, G_2)$, where G_{01} has vertex set $\{b, c\}$, a is in G_1, and e is the gate of a in G_2; then Proposition 5.41(i) establishes that c is the gate of a in G_{01} as well as in $G_{01} \cup G_2$.
(ii) Using the canonical order $\leq_0$ of G in Figure 5.8 on page 90, the meet semilattice $(V, \leq_0)$ has the set $J = abdf$ of join irreducibles, whence Proposition 5.41(ii) establishes the correspondence $\Psi_1 \leftrightarrow a$, $\Psi_2 \leftrightarrow b$, $\Psi_3 \leftrightarrow d$, $\Psi_4 \leftrightarrow f$. Using the canonical order $\leq_b$ of G in Figure 5.9, the meet semilattice $(V, \leq_b)$ has the set $J = 0cdf$ of join irreducibles, whence Proposition 5.41(ii) establishes the correspondence $\Psi_1 \leftrightarrow c$, $\Psi_2 \leftrightarrow 0$, $\Psi_3 \leftrightarrow d$, $\Psi_4 \leftrightarrow f$. $\blacksquare$

Let $C : V^* \longrightarrow 2^V \setminus \{\emptyset\}$ be a consensus rule on a lower distributive semilattice $L = (V, \leq)$. We simplify notation when no confusion arises.

Definition 5.43. *For all $P, P' \in V^*$,*

shorten	*to*
$C(P)$	CP
unique element covered by j	$p(j)$
concatenation of P and P'	PP'.

Consider what axioms would be suitable to describe such consensus rules on L. The formulations in Table 5.5 are order-theoretic analogues of corresponding axioms for hierarchies in Table 4.3 on page 73. The restated condorcet and optimality axioms use the join irreducible concept (Definition 5.9 on page 82); for all $v \in V$ and $P = (v_1, \ldots, v_k) \in V^*$, the *index of v in P* becomes

$$\gamma(v, P) = \frac{|\{i : v \leq v_i\}|}{k}.$$

Table 5.5. *Axioms: Rules on Meet Semilattices* II. *For notation see Definitions* 5.9 *on page* 82 *and* 5.43 *on the preceding page.*

***Cnd*: Condorcet** $(\forall P \in V^*)(\forall j \in J)(\forall v \in V)$
$([\gamma(j, P) = \frac{1}{2}$ and $v \vee j$ exists$] \Longrightarrow [v \vee p(j) \in CP \Longleftrightarrow v \vee j \in CP])$

***Css*: Consistency** $(\forall P, P' \in V^*)(CP \cap CP' \neq \emptyset \Longrightarrow C(PP') = CP \cap CP')$

***Fth*: Faithfulness** $(\forall P = (v_1) \in V^1)(CP = \{v_1\})$

***Opt*: Optimality** $(\forall P \in V^*)(CP \subseteq \{v \in V : \sigma_P(v)$ is minimum$\})$, where
$J(P) = \{j \in J : \gamma(j, P) \geq \frac{1}{2}\}$, $J(v) = \{j \in J : j \leq v\}$ and

$$\sigma_P(v) = \sum_{j \in J(P) \setminus J(v)} k(2\gamma(j, P) - 1)$$

By exploiting the correspondence between median graphs and median semilattices, one can use Theorem 5.35 on page 93 to characterize the median rule on median semilattices.

Theorem 5.44. McMorris, Mulder, and Powers [264, p. 229]. *Let* $C : V^* \longrightarrow 2^V \setminus \{\emptyset\}$ *be a consensus rule on a median semilattice* $L = (V, \leq)$. $C = Med$ *if and only if* C *satisfies Cnd, Css, and Fth.*

Proof. Let $C = Med_{\leq}$, the median rule relative to the standard lattice metric on $L = (V, \leq)$. For $G = (V, E)$ the covering graph of L, let Med_G be the median rule on G relative to the standard geodesic metric on G. Since the standard metrics on L and G coincide, $Med_{\leq} = Med_G$; since Med_G satisfies *Css* and *Fth* on G, so does $Med_{\leq}$ satisfy *Css* and *Fth* on L. It remains to show that $Med_{\leq}$ satisfies *Cnd* on L. Let $P \in V^*$, let j be any join irreducible of L, and suppose $\gamma(j, P) = \frac{1}{2}$. Let (G_1, G_2) be the split of G defined by edge $\{j, p(j)\}$, where $p(j) \in G_1$ and $j \in G_2$. If z in G_1 is the zero of L, then by Proposition 5.41(ii), j is the gate of z in G_2, so G_2 includes all and only the vertices $v \in V$ with $j \leq v$ in L, so that $|P_2| = |P| \cdot \gamma(j, P) = |P|/2 = |P_1|$, whence (G_1, G_2) is equal relative to P. For all $v_1 v_2 \in F_{12}$, Proposition 5.29(i) ensures that $v_1 \in Med_G P \Longleftrightarrow v_2 \in Med_G P$. Now consider any $u \in L$ for which $u \vee j$ exists. If $u \in G_2$, then $u = u \vee j = u \vee p(j)$, so trivially $u \vee p(j) \in Med_G P \Longleftrightarrow u \vee j \in Med_G P$. Let $u \in G_1$. Since $u \vee j \in G_2$ and $u \in I(z, u \vee j)$, then $I(u, u \vee j) \subseteq I(z, u \vee j)$. Let v_2 be the gate of u in G_2; since $v_2 \in I(u, u \vee j)$, then $u \leq v_2 \leq u \vee j$, but also $j \leq v_2 \leq u \vee j$, so $v_2 = u \vee j$. Since $v_1 = u \vee p(j)$ by a similar argument, then $u \vee p(j) \in Med_G P \Longleftrightarrow u \vee j \in Med_G P$, whence $Med_{\leq}$ satisfies *Cnd* on L.

Conversely, let consensus rule $C : V^* \longrightarrow 2^V \setminus \{\emptyset\}$ on $L = (V, \leq)$ satisfy *Cnd*, *Css*, and *Fth*. Clearly C satisfies *Css* and *Fth* on the covering graph $G = (V, E)$. It remains to show that C satisfies *Cnd* on G, for then $C = Med_G$ by Theorem 5.35 on page 93, whence $C = Med_{\leq}$ by Proposition 5.38 on page 94. Let $P \in V^*$ with (G_1, G_2) an equal split of G relative to P. Let the zero z of L be in G_1. By Proposition 5.41(ii), the unique join irreducible j in G_{02} is the gate of z in G_2. By Proposition 5.41(i), $p(j)$ is the gate of z in G_{01}

as well as in $G_{01} \cup G_2$. Let $uv \in F_{12}$ with $u \in G_1$; then $u = u \vee p(j)$ and $v = u \vee j$. Since C satisfies *Cnd* on L, then $u \vee p(j) \in CP \Longleftrightarrow u \vee j \in CP$, so $u \in CP \Longleftrightarrow v \in CP$, whence C satisfies *Cnd* on G. □

Theorem 5.44 on the preceding page improves on results by Barthélemy and Janowitz [46] and McMorris and Powers [270]; Theorem 4.52 on page 74, concerning the median rule on $\mathcal{H}$, follows as a corollary. McMorris and his colleagues continue to investigate the median rule on various metric spaces [261]. In view of Theorem 5.44 and since every median semilattice is lower distributive, how might the median rule be characterized on lower distributive semilattices? The problem is nontrivial since on lower distributive semilattices consensus rules exist that are faithful, consistent, and condorcet yet are not the median rule [270, p. 514]; but Theorem 4.54 on page 74 generalizes to yield the following theorem.

Theorem 5.45. McMorris, Mulder, and Powers [265, p. 6]. *Let $L = (V, \leq)$ be a lower distributive semilattice. The median rule on L is the maximum element in the set of all rules $C : V^* \longrightarrow 2^V \setminus \{\emptyset\}$ that satisfy Cnd, Css, Fth, and Opt.*

Open Problem 5.46. (B. Leclerc.) Since the median rule need not satisfy Pareto optimality on nondistributive lattices or distributive nonmedian meet semilattices, characterize those lattices or semilattices on which the median rule is Pareto optimal. ■

Open Problem 5.47. (B. Leclerc.) The complete q-quota federation consensus rules $C_q : V^* \longrightarrow V$ on a lattice $L = (V, \leq)$ satisfy a consistency axiom $(\forall P, P' \in V^*)(\forall v \in V)(C_q P = C_q P' = v \Longrightarrow C_q(PP') = v)$ in distributive and some nondistributive lattices, but not, e.g., in lattices of equivalence relations [239, §3.2]. Characterize those lattices where this type of consistency holds. ■

5.4 Remoteness Rules

Since the remoteness concept (Definition 5.24 on page 89) is the basis of the median, it can be generalized to yield other graph-theoretic consensus rules based on distance. Throughout let $G = (V, E)$ be a finite connected graph with distance $d(u, v)$ the length of a geodesic in G between $u, v \in V$.

Definition 5.48. *Let $\Re$ be the set of all real numbers. Let $R : V \times V^* \longrightarrow \Re$ be called a* remoteness function *on $G = (V, E)$. For all such R a* remoteness-based *consensus rule for G is a rule $C_R : V^* \longrightarrow 2^V \setminus \{\emptyset\}$ such that*

$$(\forall P \in V^*)(C_R(P) = \{v \in V : R(v, P) \textit{ is minimum}\}).$$

Of many possible remoteness functions these are promising and plausible.

Definition 5.49. *For all $v \in V$ and $P = (v_1, \ldots, v_k) \in V^k$ let*

$$R_1(v, P) = \sum_{i=1}^{k} d(v, v_i),$$
$$R_2(v, P) = \max\{d(v_i, v) : i = 1, \ldots, k\}, \text{ and}$$
$$R_3(v, P) = \sum_{i=1}^{k} d(v, v_i)^2.$$

$Med = C_{R_1}$ is the median rule for G. The rules based on R_2 and R_3 are named for other concepts of centrality in graphs: $Cen = C_{R_2}$ is the center rule *for G, and $Mea = C_{R_3}$ is the* mean rule *for G.*

Example 5.50. For $S = abcd$ let G be the covering graph of $(\mathcal{H}_4, \subseteq)$ (Figure 5.6 on page 83). Let $P = (H_1, H_2, H_3) \in \mathcal{H}_4^3$ with $H_1^* = \{ab, abd\}$, $H_2^* = \{ab, cd\}$, and $H_3^* = \{cd, bcd\}$. Then $Cen(P) = \{H_\emptyset, H_2\}$ and $Med(P) = \{H_2\} = Mea(P)$. ■

Although section 5.3 has elegant characterizations of *Med* in several settings, few such results are known for *Cen* or *Mea* with which to assess their usefulness.

Open Problem 5.51. Give axiomatic characterizations of *Cen* and *Mea* on the covering graph of $(\mathcal{H}_n, \subseteq)$ for hierarchies on n elements. ■

Although on $\mathcal{H}_n$ we lack axiomatic characterizations for *Mea* or *Cen*, on $\mathcal{H}_n$ the axioms *Btw* and *Css* (Table 5.4 on page 92) do distinguish among *Med*, *Mea*, and *Cen*: *Med* clearly satisfies *Btw* and *Css*; *Cen* may violate *Btw* and *Css* (Example 5.52); *Mea* may violate *Btw* (Example 5.52) but is easily seen to satisfy *Css*.

Example 5.52. For $S = abcd$ let $P = (H_1, H_3) \in \mathcal{H}_4^2$, $P' = (H_\emptyset, H_1) \in \mathcal{H}_4^2$, and $P'' = (H_\emptyset, H_3) \in \mathcal{H}_4^2$ with $H_1^* = \{abc\}$, $H_2^* = \{ab\}$, and $H_3^* = \{ab, abc\}$.

(i) Cen and Mea violate Btw since
$Cen(P'') = Mea(P'') = \{H_1, H_2\} \neq \{H_\emptyset, H_1, H_2, H_3\} = I(H_\emptyset, H_3)$.

(ii) Cen violates Css since $Cen(P) = \{H_1, H_3\}$ and $Cen(P') = \{H_\emptyset, H_1\}$, yet $Cen(PP') = Cen((H_\emptyset, H_1, H_3)) = \{H_1, H_2\} \neq \{H_1\} = Cen(P) \cap Cen(P')$. ■

On trees McMorris, Roberts, and Wang [274] characterize *Cen* in terms of population invariance (*PI* in Table 5.4 on page 92), quasi-consistency (*QC* in Table 5.4) and axioms defined only for trees. Since *PI* and *QC* make sense on any finite connected graph, they may help to characterize remoteness-based rules on other classes of graphs. *Cen* clearly satisfies *PI* in general, but by the definitions neither *Med* nor *Mea* satisfy *PI* in general, particularly not on $\mathcal{H}_n$ for interesting values of n. *Med* clearly satisfies *Css* in general, and the reader can verify that *Mea* does as well.

Proposition 5.53. *The mean rule Mea satisfies Css on any finite connected graph.*

Since *Css* implies *QC*, then *Med* and *Mea* satisfy *QC* in general. Although *Cen* need not satisfy *Css* (Example 5.52, *Cen* nevertheless satisfies *QC* in general.

Theorem 5.54. [274, p. 85] *The center rule Cen satisfies QC on any finite connected graph.*

Proof. For finite connected $G = (V, E)$ let $P, P' \in V^*$ with $Cen(P) = Cen(P')$, $v \in Cen(PP')$, and $u \in Cen(P) = Cen(P')$. From the definition of *Cen* we have $R_2(v, PP') \leq R_2(u, PP')$, $R_2(u, P) \leq R_2(v, P)$, and $R_2(u, P') \leq R_2(v, P')$. Clearly

$$\begin{aligned} R_2(u, PP') &= \max\{R_2(u, P), R_2(u, P'\} \\ &\leq \max\{R_2(v, P), R_2(v, P')\} \\ &= R_2(v, PP') \leq R_2(u, PP'). \end{aligned} \tag{5.1}$$

Thus $R_2(v, PP') = R_2(u, PP')$, which implies $u \in Cen(PP')$, whence $Cen(P) \subseteq Cen(PP')$. We claim that $v \in Cen(P) = Cen(P')$. For this it suffices to show that $R_2(v, P) = R_2(u, P)$ or $R_2(v, P') = R_2(u, P')$. From (5.1),

$$\max\{R_2(u, P), R_2(u, P')\} = \max\{R_2(v, P), R_2(v, P')\},$$

but if $R_2(u, P) < R_2(v, P)$ and $R_2(u, P') < R_2(v, P')$, then

$$\max\{R_2(u, P), R_2(u, P')\} < \max\{R_2(v, P), R_2(v, P')\},$$

a contradiction. $\square$

5.5 Notes

Basic concepts of graphs are reviewed in most standard texts on discrete mathematics, e.g., Ross and Wright [352], while Davey and Priestley [148] provide an excellent introduction to ordered sets and lattices; for advanced treatments see Harary [202] or Berge [65] for graph theory or Birkhoff [72], Crawley and Dilworth [142], or Grätzer [185] for lattice theory. The rich literature on general models of consensus includes the following papers. Mirkin [285] uses domain assumptions, independence, neutrality, and monotonicity to characterize consensus rules in terms of sets of coalitions. Barthélemy [42] establishes Arrow's impossibility theorem in a general ordinal case where some configurations are allowed in the consensus rule's domain, and that domain is included in the codomain. Leclerc [233] gives the general form of consensus rules on valued (fuzzy) quasi-orders that satisfy Arrow-like conditions of efficiency and binariness, a result that generalizes many previous consensus results. Barthélemy, Leclerc, and Monjardet [48] survey how ordered sets can be used in classification and systematics. Day, McMorris, and Meronk [154] characterize consensus rules that are based on iterated maximal lower bound operators in posets. Monjardet [296] describes an order-theoretic formalization so as to account for existence and characterization results in disparate fields where logical problems of aggregation arise. Crown, Janowitz, and Powers [144, 145, 146] use ideas of Barthélemy and Monjardet [54] to view consensus objects as the result of gluing bricks together in a suitable manner, so as to understand what it is about the bricks that produces analogues of Arrow's theorem. Cramer-Benjamin [140] investigates definitions of independence that apply to the ordinal model [144] and to much more general classes of closed set systems. Although for hierarchies the process of gluing bricks together can lead to more than one consensus object, Crown and Janowitz [143] show that one cannot get very far away from closed weak hierarchies. Cramer-Benjamin, Crown, and Janowitz [141] impose conditions on a consensus rule, rather than on the output structure, so as to reveal what makes Arrovian results hold in vastly different settings.

Sections 5.2.1 (Meet Projection Rules) and 5.2.2 (Federation Rules) are based on Monjardet's [296] elegant paper. Leclerc [235] extends Monjardet's [296] formalizations to consensus rules on lattices of valued objects such as fuzzy preorders (Leclerc [233]). Leclerc and Monjardet [240] generalize and synthesize the Arrovian formalizations of Monjardet [296] and Leclerc [235] to obtain the foundations of an axiomatic theory of consensus rules on lattices. In [235, 240, 296] the join-irreducible dependence relation δ [300] is denoted by β. Axiomatic investigations of oligarchic consensus rules, e.g., Leclerc [233], Leclerc and Monjardet [240], Stinebrickner [390], and Neumann and Norton [315], are relevant to the consensus of classification trees called dendrograms, e.g., Lapointe and Cucumel [231] and Chepoi and Fichet [127], which are equivalent to ultrametric functions (Johnson [213]).

For section 5.3 (Median Rules): Sholander [378] and Avann [22] introduce the concepts of median semilattice and median graph, respectively. See McMorris, Mulder, and Powers [264] for the median rule on median semilattices; McMorris and Powers [270] and McMorris, Mulder, and Powers [265] for the median rule on lower distributive semilattices; Bandelt and Barthélemy [27], McMorris, Mulder, and Roberts [266], and McMorris, Mulder, and Powers [264] for the median rule on graphs. Our view of median graphs is strongly influenced by Mulder's insightful papers [303, 304, 305, 306]. Leclerc [239] investigates the consequences for consensus rules (including the median rule) of the existence of semilattice structure in the set of partial orders of a finite set. Monjardet [294] surveys characterizations of metrics on various ordered sets. Since the median rule relies on a simple optimization criterion involving distances, one knows the median rule if one can characterize the distance function: for characterizations of distances on other relations or structures, see Kemeny [222] and Kemeny and Snell [224] for weak orders; Bogart [98, 99] for partial orders and asymmetric relations; Margush [245] for binary relations and hierarchies. Researchers have been intrigued by the relationships between majorities and medians in different types of ordered sets: see Barbut [35] and Monjardet [292] for distributive lattices; Barthélemy [41] for modular lattices; Leclerc [234] for semimodular lattices; Bandelt and Barthélemy [27] for median semilattices; Leclerc [236, 237] for lower distributive semilattices; and Powers [339] for semimodular posets.

For section 5.4 (Remoteness Rules): McMorris [262] reviews recent results concerned with characterizing the median, mean, and center functions on metric spaces. If consensus objects are constrained to be vertices of a graph, McMorris, Roberts, and Wang [274] characterize the center consensus rule on trees, while Biagi [66] studies the mean rule on trees. Holzman [206] and Vohra [407] characterize the mean and median rules on trees in the general case where consensus objects can be located anywhere along a graph's edges.

Chapter 6

Beyond Consensus

The holy grail of phylogenetics is the reconstruction of the one true tree of life.
— J. L. Thorley and R. D. M. Page [402, p. 486]

[T]he use of composite phylogenies to study evolutionary patterns is one of the few choices left to the investigators. However, choosing between the various supertree methods is not a straightforward task.
— N. Salamin, T. R. Hodkinson, and V. Savolainen [363, p. 136]

An unspoken sentiment is that supertree construction merely represents a stopgap measure ... to infer the tree of life until there is sufficient molecular data.
— O. R. P. Bininda-Emonds, J. L. Gittleman, and M. A. Steel [70, p. 283]

Circumstances may arise in which consensus rules modelled by the functions $C : \mathcal{X}^k \longrightarrow \mathcal{X}$ or $C : \mathcal{X}^* \longrightarrow \mathcal{X}$ are inappropriate, inadequate, or irrelevant. Consider these examples.

Example 6.1. Agreement in Theory. Let $S = abcde$ and consider (Figure 6.1) the profile $P = (H_1, H_2) \in \mathcal{H}^2$, where $H_1^* = \{ab, abc, abcd\}$ and $H_2^* = \{ae, abe, abce\}$. Then $Str P = Maj P = H_\emptyset$ even though H_1 and H_2 agree except for the position at which leaf e is attached in hierarchy H_3 on $S' = abcd$. By ignoring the effects of the root vertices in H_1–H_3, we obtain an analogous problem on phylogenies (Figure 6.2): the structure shared by phylogenies T_1, T_2 on $S = abcde$ is better represented by the phylogeny T_3 on $S' = abcd$ than by any phylogeny on S. ■

Example 6.2. Agreement in Practice. Let $S = \{1, \ldots, 11\}$ and consider (Figure 6.3 on page 105) the treelike structures which, after deletion of a suspected hybridization, could be modeled by hierarchies with at most ten labels. "The significance of the identity of the two reduced area-cladograms is not the fact that they each include the same areas, but that each includes the *same* areas in the *same* cladistic sequence." [351, pp. 179]. ■

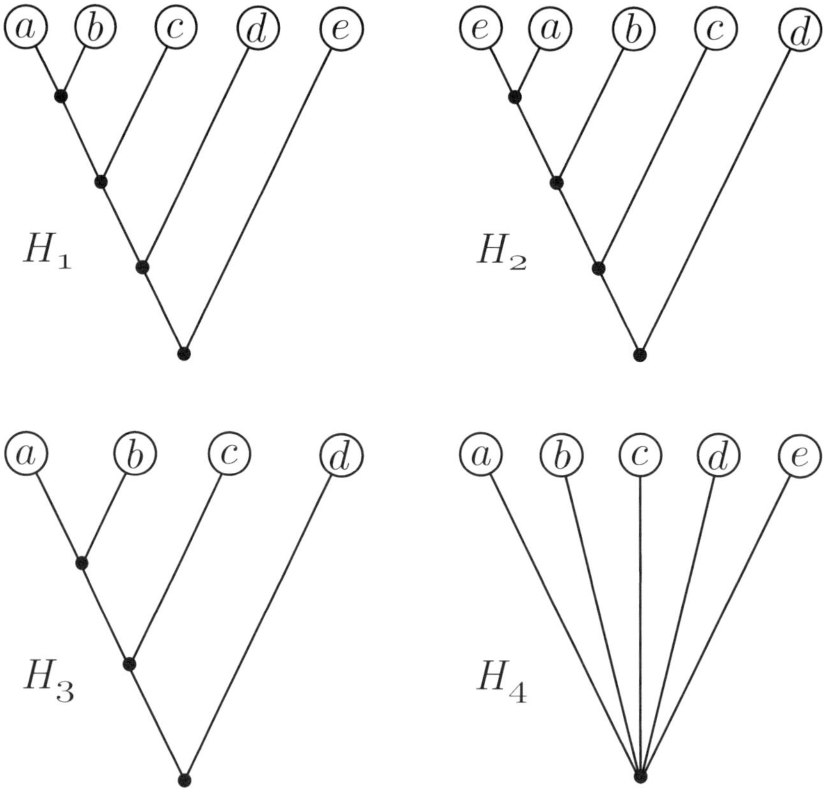

Figure 6.1. *Agreement Problem for Hierarchies. Example* 6.1 *explains.*

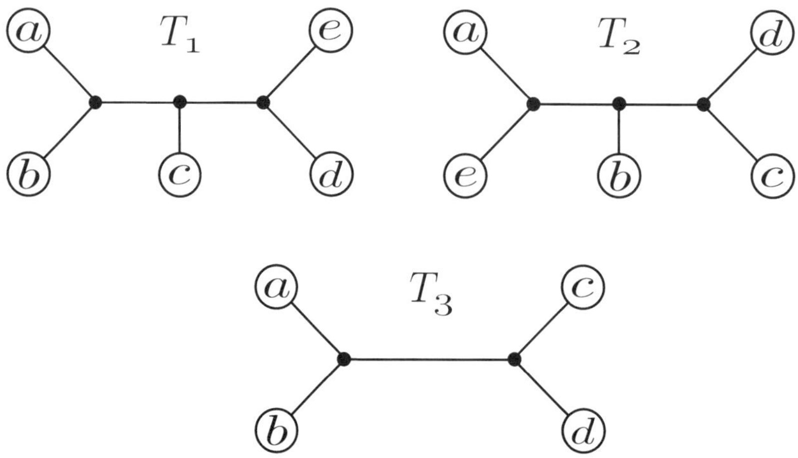

Figure 6.2. *Agreement Problem for Phylogenies. Example* 6.1 *explains.*

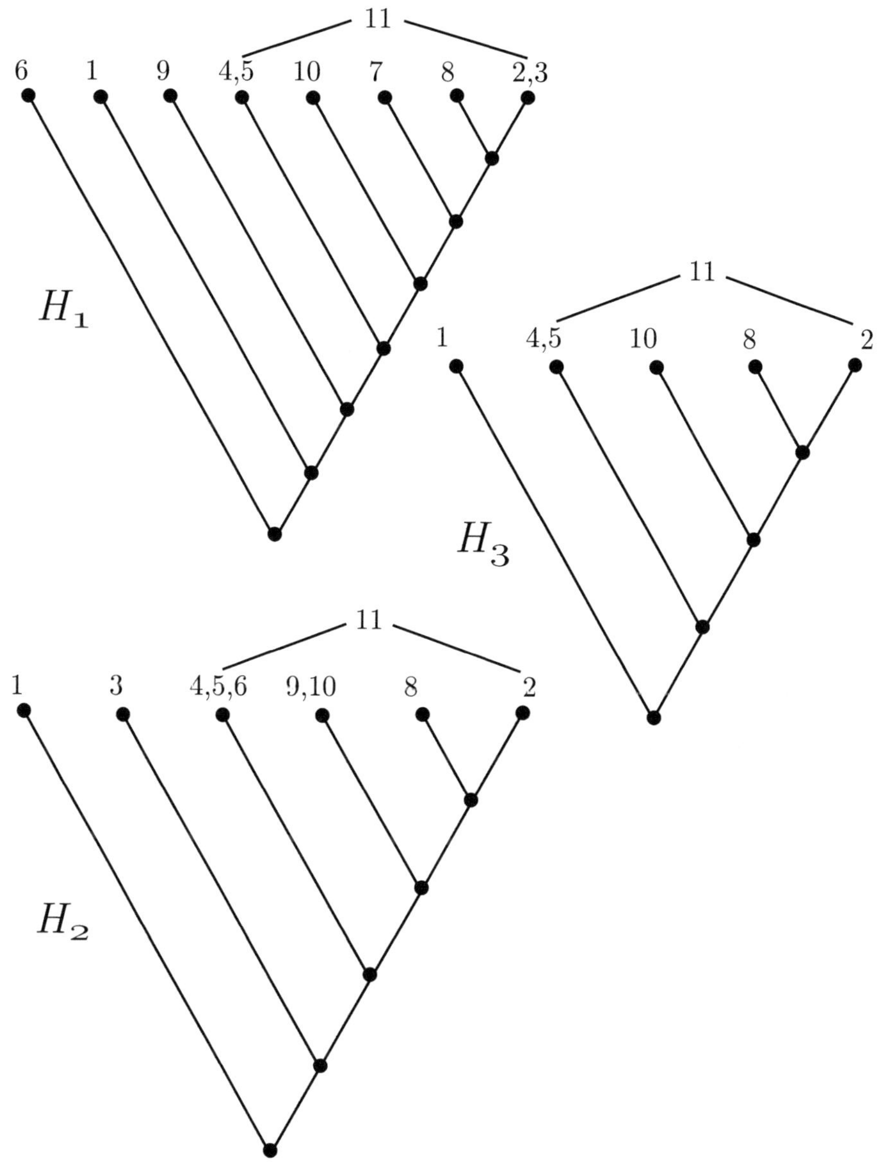

Figure 6.3. *Congruence of Area Cladograms for Poeciliid Fish* [351, pp. 179–181]. H_1: *Area cladogram of Middle American* Heterandria. H_2: *Simplified area cladogram of Middle American* Xiphophorus. H_3: *Cladogram representing residual congruence, obtained from* H_1 *and* H_2 *by deleting unique* (7) *and incongruent* (3, 6, 9) *elements. Area* 11 *denotes a suspected hybridization.*

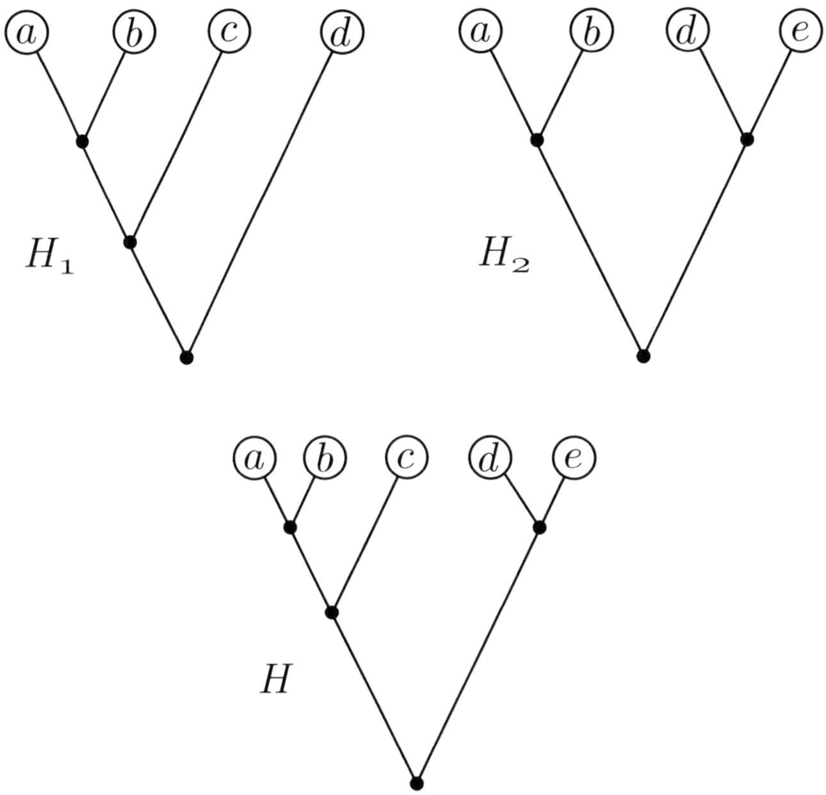

Figure 6.4. *Synthesis Problem for Hierarchies. Example* 6.3 *explains.*

Example 6.3. Synthesis. Let $S = abcde$ and consider (Figure 6.4) the profile $P = (H_1, H_2)$, where H_1 and H_2 are hierarchies on $S_1 = abcd$ and $S_2 = abde$, respectively. Since each cluster in H_1^* is compatible with every cluster in H_2^*, we may construct a hierarchy H on S with $H^* = H_1^* \cup H_2^*$, where H perfectly represents P in the sense that it has all and only the nontrivial clusters of the profile's hierarchies. By ignoring the effects of the root vertices in H_1 and H_2, we obtain an analogous problem on phylogenies (Figure 6.5), where we may construct a phylogeny T_3 on $S = abcde$ whose quartet representation $q(T_3)$ includes both $q(T_1)$ and $q(T_2)$. ■

Although our consensus rules on $\mathcal{H}$ or $\mathcal{P}$ have been defined for profiles P where each member of P has the same leaf set S, Example 6.2 on page 103 and Example 6.3 show that it is reasonable to relax this requirement, while Examples 6.1 and 6.2 on page 103 show that it is reasonable to let aggregation rules return a result that is defined on a proper subset of S. We will investigate agreement (subtree) and synthesis (supertree) problems for phylogenies and hierarchies since such structures are relevant to systematic biology. Although much research has concerned algorithms for subtree or supertree construction, we will focus on axiomatic considerations where much less is known.

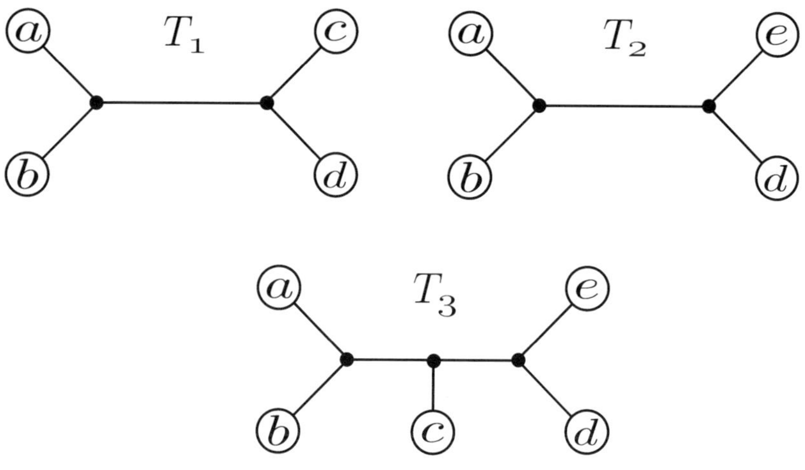

Figure 6.5. *Synthesis Problem for Phylogenies. Example* 6.3 *explains.*

6.1 Phylogenies

For $T \in \mathcal{P}$ and $X \subseteq S$, we defined the restriction $T|_X$ in Definition 3.29 on page 38. $T|_X$ can be viewed as the phylogeny that results by removing (pruning) all the leaves of T that are labeled by members of $S \setminus X$ and suppressing any vertices of degree 2 that result from doing so. Of course, if $|X| = 4$, then $T|_X$ is simply a quartet. Now for $X \subseteq S$ and $|X| \geq 4$, let $\mathcal{P}|_X$ be the set of all phylogenies on X, and let $\mathcal{P}_s = \bigcup_{X \subseteq S} \mathcal{P}|_X$. Analogous to the consensus rules $C : \mathcal{P}^k \longrightarrow \mathcal{P}$, an *agreement* or *subtree rule* on $\mathcal{P}$ is a function

$$C : \mathcal{P}^k \longrightarrow \mathcal{P}_s,$$

while a *synthesis* or *supertree rule* on $\mathcal{P}$ is a function

$$C : \mathcal{P}_s^k \longrightarrow \mathcal{P}.$$

To study subtree or supertree rules, consider relationships between phylogenies that depend on whether and how one phylogeny can be transformed into another.

Definition 6.4. *For $T, T' \in \mathcal{P}$, T is said to* resolve *T' if T' can be obtained from T by contracting edges. T is said to* display *$T'|_X$ if $T|_X = T'|_X$ or $T|_X$ resolves $T'|_X$. A profile $P = (T_1, \ldots, T_k)$ is said to* display *$T'|_X$ if $T'|_X$ is displayed by each T_i, $i = 1, \ldots, k$. Let $D(P)$ be the set of all nontrivial phylogenies (i.e., those having at least one resolved quartet) that are displayed by P. A set $\{T_1, \ldots, T_m\}$ of phylogenies is called* compatible *if there is a phylogeny that displays each T_i.*

Thus T displays T' if there is a sequence of leaf deletions (the restricting part) and edge contractions (the resolving part) by which T can be transformed into T'.

Table 6.1. *Axioms: Rules on Phylogenies* II. *For notation see Definition* 6.6.

***Agr*: Agreement**	$(\forall P \in \mathcal{P}^k)(D(P) \neq \emptyset \Longrightarrow CP \in D(P))$
***Dsp*: Display**	$(\forall P \in \mathcal{P}^k)(\{P\}$ compatible $\Longrightarrow (\forall i \in K)(CP$ displays $T_i))$
***S-Ntr*: *S*-Neutrality**	$(\forall P \in \mathcal{P}^k)(\forall$ permutations ϕ of $S)(C(\phi(P)) = \phi(CP))$
***PO*: Pareto Optimality**	$(\forall w, x, y, z \in S)(\forall P \in \mathcal{P}^k)(wxT_K yz \Longrightarrow wxCPyz)$
***Sym*: Symmetry**	$(\forall P \in \mathcal{P}^k)(\forall$ permutations σ of $K)(CP = C(T_{\sigma(1)}, \ldots, T_{\sigma(k)}))$

Example 6.5. In Figure 6.2 on page 104, since both T_1 and T_2 display T_3, then the profile $P = (T_1, T_2)$ displays T_3 and in fact $D(P) = \{T_3\}$. In Figure 6.5, since T_3 displays both T_1 and T_2, then the set $\{T_1, T_2\}$ is compatible. ■

Phylogenies also can be transformed by relabeling their leaves. If ϕ is a permutation of S and $T \in \mathcal{P}$, then $\phi(T)$ is the phylogeny obtained from T by replacing each $x \in S$ with $\phi(x)$. If $P = (T_1, \ldots, T_k) \in \mathcal{P}^k$, then $\phi(P) = (\phi(T_1), \ldots, \phi(T_k))$. And when no confusion arises we will simplify notation as follows.

Definition 6.6. *For all* $w, x, y, z \in S$ *and* $P = (T_1, \ldots, T_k) \in \mathcal{P}^k$,

shorten	*to*
$C(P)$	CP
$q(T_i)$	T_i
$wx\|yz \in q(T_i)$	$wxT_i yz$
$(\forall i \in I \subseteq K)(wxT_i yz)$	$wxT_I yz$
set of phylogenies in P	$\{P\}$.

Consider what axioms would be suitable to describe rules for the subtree, consensus, or supertree problems on phylogenies. The formulations in Table 6.1 include symmetry (as in Table 5.4 on page 92), Pareto optimality (as for phylogenies in Table 3.3 on page 39), and a variant of neutrality, which asserts that if ϕ is used to relabel the leaves of phylogenies in P, it makes no difference whether the relabeling occurs before $(C(\phi(P)))$ or after $(\phi(CP))$ C is applied. Two axioms are new and address basic aspects of agreement or synthesis. For subtree rules, the agreement axiom (*Agr*) asserts that if at least one nontrivial phylogeny is displayed by P, then CP must have that desirable property. For supertree rules, the display axiom (*Dsp*) asserts that if at least one phylogeny displays every $T_i \in P$, then CP must have that desirable property. But as natural as these axioms may be, their use in combination nevertheless leads to three impossibility results described below. It is remarkable that each proof hinges on the existence of two phylogenies (Figure 6.6). For subtree rules our impossibility result is the following theorem.

Theorem 6.7. *No subtree rule on phylogenies with six or more leaves can satisfy Agr and S-Ntr.*

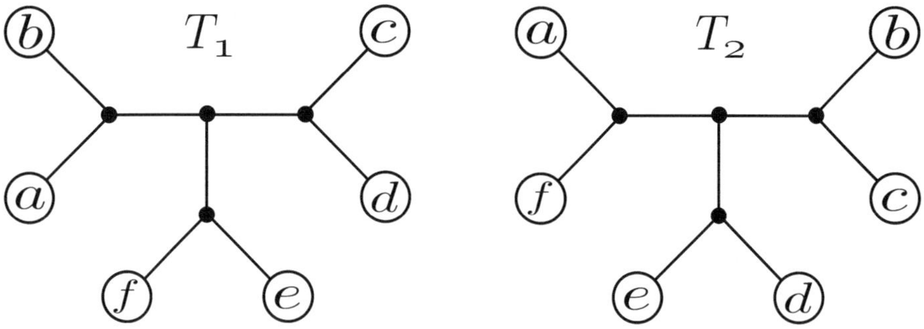

Figure 6.6. *Problematic Phylogenies* [97, pp. 5–6]. *T_1 and T_2 are the only phylogenies on $S = abcdef$ that display the quartets $ab|de$, $af|cd$, and $bc|ef$.*

Proof. Consider $P = (T_1, T_2)$ with T_1 and T_2 as in Figure 6.6. Readers can verify that $D(P)$ consists of $ab|de$, $af|cd$, and $bc|ef$. Then *Agr* requires that $CP \in \{ab|de, af|cd, bc|ef\}$. Consider the permutation $\phi = (aec)(bfd)$. Then $\phi(T_1) = T_1$ and $\phi(T_2) = T_2$, whence $\phi(P) = P$. Now *S-Ntr* requires that $C(\phi(P)) = \phi(CP)$, so that $CP = \phi(CP)$; but this is impossible since $\phi(ab|de) = bc|ef$, $\phi(bc|ef) = af|cd$, and $\phi(af|cd) = ab|de$. □

For consensus rules the next result can be compared with Theorem 3.32 on page 39.

Theorem 6.8. Steel, Dress, and Böcker [386, p. 366]. *No consensus rule on phylogenies with six or more leaves can satisfy PO, S-Ntr, and Sym.*

Proof. Consider $P = (T_1, T_2)$, where T_1 and T_2 are in Figure 6.6. We have seen that T_1 and T_2 both display $ab|de$, $af|cd$, and $bc|ef$. Then *PO* requires that CP display $ab|de$, $af|cd$, and $bc|ef$. Böcker, Dress, and Steel [97] established that T_1 and T_2 are the only phylogenies to display these three quartets simultaneously, whence $CP \in \{T_1, T_2\}$. Consider the permutation $\phi = (bf)(ce)$. Notice that $\phi(T_1) = T_2$ and $\phi(T_2) = T_1$, whence *Sym* ensures that $C(\phi(P)) = CP$. Now *S-Ntr* requires that $C(\phi(P)) = \phi(CP)$, so that $CP = \phi(CP)$. But this is impossible since $\phi(T_1) \neq T_1$ and $\phi(T_2) \neq T_2$. □

Since nothing in the proof of Theorem 6.8 requires that the consensus rule's codomain be $\mathcal{P}$, rather than $\mathcal{P}_s$, the result also holds for subtree rules. As for supertree rules, the impossibility result is as follows.

Theorem 6.9. Steel, Dress, and Böcker [386, p. 364]. *No supertree rule on phylogenies with six or more leaves can satisfy Dsp, S-Ntr, and Sym.*

Proof. Let $S = abcdef$ and consider $P = (ab|de, af|cd, bc|ef)$. These phylogenies form a compatible set. T_1 and T_2 in Figure 6.6 are the only phylogenies to display these three quartets simultaneously (Böcker, Dress, and Steel [97, p. 5]), whence *Dsp* requires that $CP \in \{T_1, T_2\}$. Consider the permutation $\phi = (bf)(ce)$. Notice that $\phi(ab|de) = af|cd$,

$\phi(af|cd) = ab|de$, and $\phi(bc|ef) = bc|ef$, whence *Sym* ensures that $C(\phi(P)) = CP$. Now *S*-*Ntr* requires that $C(\phi(P)) = \phi(CP)$, so that $CP = \phi(CP)$. But this is impossible since $\phi(T_1) \neq T_1$ and $\phi(T_2) \neq T_2$. □

Open Problem 6.10. Precisely how strong is S-neutrality? For agreement, consensus, or synthesis rules on phylogenies, characterize those that satisfy *S*-*Ntr*. ■

6.2 Hierarchies

Since agreement, consensus, and synthesis rules on phylogenies fail to satisfy reasonably desirable properties, consider the situation for hierarchies. It is easy to give the analogous definitions of subtree and supertree rules on hierarchies. Notice that Theorems 6.8 and 6.9 do not hold for hierarchies: the Adams consensus rule on $\mathcal{H}$ satisfies *Sym*, *S*-*Ntr*, and *TPO*, and the MINCUTSUPERTREE rule (Semple and Steel [369]) on $\mathcal{H}$ satisfies *Dsp*, *S*-*Ntr*, and *Sym*. However, we do not know if Theorem 6.7 on page 108 holds for hierarchies or if there is a subtree rule on $\mathcal{H}$ that satisfies *Agr* and *S*-*Ntr*.

Open Problem 6.11. Does there exist a subtree rule on $\mathcal{H}$ that satisfies *Agr* and *S*-*Ntr*? ■

The MINCUTSUPERTREE synthesis rule is related to the Adams consensus rule when all hierarchies are defined on the same leaf set.

Open Problem 6.12. Semple and Steel [369] have investigated the relationship between the Adams consensus result $C_a P$ for a profile P of hierarchies on the same leaf set and the supertree returned by the MINCUTSUPERTREE algorithm when it is applied to P. They ask if MINCUTSUPERTREE could be modified so that, when applied to hierarchies on the same leaf set, the Adams consensus is equal to the corresponding supertree or can be obtained by contraction from an induced subtree of it [369, p. 157]. If so, could the modified algorithm be characterized axiomatically so that, when restricted to profiles of hierarchies on the same leaf set, it yields a new characterization of the Adams consensus rule? ■

6.3 Notes

For the agreement problem: Rosen [351, pp. 179–182] and Gordon [182] first proposed using agreement subtrees to represent the information shared by two trees, an *agreement subtree* of $P = (T_1, \ldots, T_k)$ being a tree T where $T = T_1|_X = T_2|_X = \cdots = T_k|_X$ for some $X \subseteq S$. A common goal in this area is to find agreement subtrees for P with the largest number of leaves. Gordon [182] and Finden and Gordon [162] describe heuristic algorithms to calculate agreement subtrees, which they call common pruned trees. This early work stimulated much algorithmic research, e.g., Steel and Warnow [387], Goddard et al. [179], Keselman and Amir [225], Kubicka, Kubicki, and McMorris [228], Farach, Przytycka, and Thorup [159], Amir and Keselman [6], Bryant [112], Przytycka [340], Gupta and Nishimura [194], Kao [217], and Cole et al. [135]. Kubicka, Kubicki, and McMorris [227], Goddard and Kubicki [181], and Bryant, McKenzie, and Steel [114] investigate the structure, size and/or number of agreement subtrees. Metrics based on the agreement subtree of two trees are developed by Goddard et al. [179, 180] and Kubicka, Kubicki, and McMorris [229]. Gordon [182] describes a consensus rule that regrafts vertices on an agreement subtree; Finden and Gordon [162] show that this pruned-and-regrafted consensus rule is a refinement of the strict consensus rule; Goddard et al. [180] begin an axiomatic study of such rules. Wilkinson [413, 414, 415] addresses problems of insensitivity and ambiguity in the strict and Adams consensus rules by developing new strict and semistrict consensus rules based on obtaining reduced subtrees by pruning leaves; this research is reviewed and extended by Wilkinson and Thorley [417].

For the supertree problem: In 1986 Gordon [183] introduced the supertree problem on profiles of two hierarchies and proposed algorithms for its solution. Many algorithms solving the supertree problem for hierarchies have since been developed, e.g., Steel [385], Lanyon [230], Constantinescu and Sankoff [138], Ng and Wormald [316], Strimmer and von Haeseler [394], Henzinger, King, and Warnow [204], Böcker et al. [96], and Semple and Steel [369]. Wilkinson et al. [418] classify supertree rules by the types of tree representation and analysis: MRC (matrix representation with compatibility), e.g., Purvis [341], Rodrigo [349], and Pisani [329]; MRD (matrix representation with distances), e.g., Lapointe and Cucumel [231]; MRF (matrix representation with flipping), e.g., Chen et al. [126]; MRP (matrix representation with parsimony), e.g., Baum [61], Ragan [342], Baum and Ragan [62], Ronquist [350], Bininda-Emonds and Bryant [69], Bininda-Emonds and Sanderson [71], Pisani and Wilkinson [330], and Bininda-Emonds [67]. Wilkinson and Thorley [416] and Thorley and Wilkinson [403] apply reduced consensus concepts in the construction of supertrees. Steel, Dress, and Böcker [386] use an axiomatic approach to explore the theoretical limitations of supertree and consensus rules, while Wilkinson et al. [418] informally describe axioms for assessing the relevance of supertree methods to biological applications. For the biological relevance of the supertree problem see, e.g., Sanderson, Purvis, and Henze [364], Soltis and Soltis [384], Salamin, Hodkinson, and Savolainen [363], Bininda-Emonds, Gittleman, and Steel [70], and Pisani et al. [331]. Bininda-Emonds [68] presents an up-to-date collection of papers on types, applications, examples, methodologies, and criticisms of supertrees.

The field of science, indeed, the whole world of human society, is a cooperative one. At each moment, we are competing, whether for academic honors or business success. But the background, and what makes society an engine of progress, is a whole set of successes and even failures from which we all have learned.
— K. J. Arrow [19, p. 57]

Appendix A
Quick References

Table A.1 shows numbered conventions that are scattered throughout the text.

Table A.2 brings together the various notational conventions we have used.

Table A.3 is an index to the tables (settings) in which each axiom is defined.

Tables A.4–A.5 give the open problems that are scattered throughout the text.

Table A.1. *Conventions*

No.	Description
1.4	Unless specifically stated otherwise, the set K of individuals is finite with $\|K\| = k \geq 2$.
1.5	If $X = \{a, b, \ldots, y, z\}$ is any set of elements denoted by single letters (or digits), we may shorten its representation to $X = ab \cdots yz$. For example, $Y = \{\{a, c, d\}, \{a, f\}, \{b, d, f, g\}, \{e\}\} = \{acd, af, bdfg, e\}$. Also we may shorten any ordered pair (a, b) to ab. Context will determine whether ab means $\{a, b\}$ or (a, b).
1.7	To reduce the paired parentheses required to specify the meaning of logical sentences, we apply logical operators in the order first $\neg$, then $\vee$ and $\wedge$, finally $\Longrightarrow$ and $\Longleftrightarrow$; thus $\neg p \wedge q$ means $(\neg p) \wedge q$ rather than $\neg(p \wedge q)$. Parentheses always delimit the formula to which quantifiers apply; thus in $(\exists x)(x < y) \vee y = 0$ the quantifier applies only to $(x < y)$.
2.4	To any (multi)consensus rule C with domain $\mathcal{O}^k$ or $\mathcal{E}^k$ or $\mathcal{T}^k$ is associated a set $S = S_n$ of n alternatives on which the relations are defined. In this context, unless specifically stated otherwise, S is finite with $\|S\| = n \geq 3$.
3.1	In Chapter 3 and thereafter, unless specifically stated otherwise, each (multi)consensus rule C is assumed to be collectively rational, i.e., $C(P)$ is defined and single valued for every profile P.
3.26	In the biological literature a phylogeny may be either unrooted or rooted. In this book, unless otherwise qualified, a phylogeny will be unrooted, and the term hierarchy will denote the rooted phylogeny of the biologist.
3.30	To any (multi)consensus rule C with domain $\mathcal{P}^k$ is associated a set $S = S_n$ of n leaf labels on which the phylogenies are defined. In this context, unless specifically stated otherwise, S is finite with $\|S\| = n \geq 5$.
3.45	To any (multi)consensus rule C with domain $\mathcal{H}^k$ or $\mathcal{W}^k$ is associated a set $S = S_n$ of n leaf labels on which the hierarchies are defined. In this context, unless specifically stated otherwise, S is finite with $\|S\| = n \geq 5$.
3.65	To any (multi)consensus rule C with domain $\mathcal{W}_c^k$ is associated a set $S = S_n$ of n leaf labels on which the hierarchies are defined. In this context, unless specifically stated otherwise, S is finite with $\|S\| = n \geq 6$.

Table A.2. *Notations*

This	Means This	Occurs in Definition
CQ or R	$C(Q)$	2.3 on page 13
$N(xPy)$	$\lvert\{i \in K : xP_iy\}\rvert$	2.3 on page 13
$N(xIy)$	$\lvert\{i \in K : xI_iy\}\rvert$	2.3 on page 13
$N(xRy)$	$\lvert\{i \in K : xR_iy\}\rvert$	2.3 on page 13
xP_Iy	$(\forall i \in I \subseteq K)(xP_iy)$	2.3 on page 13
$I_{xy}(Q)$	$\{i \in I \subseteq K : xR_iy\}$	2.3 on page 13
$K_{xy}(Q)$	$\{i \in K : xR_iy\}$	2.3 on page 13
CP	$C(P)$	3.6 on page 29, and elsewhere
xE_Iy	$(\forall i \in I \subseteq K)(xE_iy)$	3.6 on page 29
$xR^\star_Iy$	$(\forall i \in I \subseteq K)(xR^\star_iy)$	3.19 on page 35
T_i	$q(T_i)$	3.31 on page 38, 6.6 on page 108
wxT_iyz	$wx\|yz \in q(T_i)$	3.31 on page 38, 6.6 on page 108
$wxyzT_i$	$wxyz \in q(T_i)$	3.31 on page 38
wxT_Iyz	$(\forall i \in I \subseteq K)(wxT_iyz)$	3.31 on page 38, 6.6 on page 108
$wxyzT_I$	$(\forall i \subset I \subseteq K)(wxyzT_i)$	3.31 on page 38
$X \in H_I$	$(\forall i \in I \subseteq K)(X \in H_i)$	3.49 on page 44
$I_X(P)$	$\{i \in I \subseteq K : X \in H_i\}$	3.49 on page 44
$K_X(P)$	$\{i \in K : X \in H_i\}$	3.49 on page 44
R_i	$t(H_i)$	3.49 on page 44
R	$t(CP)$	3.49 on page 44
xyR_iz	$xy\|z \in t(H_i)$	3.49 on page 44
$xyzR_i$	$xyz \in t(H_i)$	3.49 on page 44
xyR_Iz	$(\forall i \in I \subseteq K)(xyR_iz)$	3.49 on page 44
$(x)_k$	$(x, \ldots, x) \in L^k$	5.15 on page 85
$K_j(P)$	$\{i \in K : j \leq x_i\}$	5.15 on page 85
PP'	concatenation of P and P'	5.30 on page 92, 5.43 on page 96
$\{P\}$	set of elements in P	5.30 on page 92, 6.6 on page 108
$p(j)$	unique object covered by j	5.43 on page 96

Table A.3. *Axioms: Index*

Abbr.	Name	Defined on Page
Agr	Agreement	108
APO	Anti-Pareto Optimality	14
Atn	Autonomy	14, 54, 86
Btw	Betweenness	62, 92
Cnd	Condorcet	73, 92, 97
CPO	co-Pareto Optimality	54
CR	Collective Rationality	14
Css	Consistency	73, 92, 97
Cst	Constant	14, 29, 39, 86
Dct	Dictatorship	14, 29, 35, 39, 45, 54
DM	Decisive Monotonicity	54, 86
DN	Decisive Neutrality	14, 54, 62, 86
Dsp	Display	108
Eff	Efficiency	73
Ext	Extensiveness	86
FT	Free Triples	14
Fth	Faithfulness	73, 92, 97
ID	Inverse Dictatorship	14
Idm	Idempotence	86
Ind	Independence	14, 29, 35, 39, 45, 54, 62, 73, 86
Ist	Isotony	86
MC	Meet Compatibility	86
MN	Monotonic Neutrality	54, 86
NP	Nesting Preservation	62
Ntr	Neutrality (see also *S-Ntr*)	86
Olg	Oligarchy	29
Opt	Optimality	73, 97
PI	Population Invariance	92
PO	Pareto Optimality	14, 29, 35, 39, 45, 54, 62, 108
PR	Positive Responsiveness	14
Prj	Projection	29, 39, 45, 62
QC	Quasi-Consistency	92
QSP	Qualified Strong Presence	62
RI	Removal Independence	45
RTI	Removal Ternary Independence	45
S-Ntr	*S*-Neutrality	108
SP	Strong Presence	62
Sym	Symmetry	14, 29, 45, 54, 73, 86, 92, 108
TPO	Ternary Pareto Optimality	45
Unn	Unanimity	86
USP	Upper Strong Presence	62
WI	Weak Independence	45

Table A.4. *Open Problems* I

No.	Description
3.16	(M. F. Janowitz.) When *Ind* and *PO* hold for consensus rules on weak orders, then dictators are weak; when they hold for rules on equivalence relations, then dictators are strong. How does the type of relation determine whether dictators are weak or strong?
3.24	Characterize consensus rules on tree quasi-orders that satisfy *Ind*.
3.39	Since *Ind* and *PO* appear to be so strong for consensus rules on phylogenies, how might they be weakened so as to obtain characterizations of rules on phylogenies that are perhaps more relevant than projections?
3.71	Determine if other reasonable concepts of independence or Pareto optimality give rise to impossibility theorems for consensus rules on $\mathcal{W}$ or $\mathcal{W}_c$.
4.11	Let similarity or dissimilarity data be given for a study collection S and let a set K of clustering programs be used to analyze these data. For $L \subseteq K$ let the $l = \|L\|$ algorithms indexed by L yield weak hierarchies as output, while those indexed by $L^c = K \setminus L$ yield strong hierarchies. Describe rigorously whatever is in common agreement among these k hierarchies.
4.33	Characterize consensus rules on hierarchies that satisfy *Btw*.
4.55	Use conceptually simple properties to characterize the median rule on $\mathcal{W}$.
4.56	[272, p. 268]. Characterize the set of complete multiconsensus rules on $\mathcal{W}$ that satisfy *Cnd*, *Css*, and *Fth*.

Table A.5. *Open Problems* II

No.	Description
5.32	Give an example of a consensus rule on a median graph that satisfies *Btw*, *Css*, and *Sym* but is not the median rule.
5.33	Do there exist other classes of graphs (or metric spaces) for which the median rule can be characterized by *Btw*, *Css*, and *Sym*?
5.46	(B. Leclerc.) Since the median rule need not satisfy Pareto optimality on nondistributive lattices or distributive nonmedian meet semilattices, characterize those lattices or semilattices on which the median rule is Pareto optimal.
5.47	(B. Leclerc.) The complete q-quota federation consensus rules $C_q : V^* \longrightarrow V$ on a lattice $L = (V, \leq)$ satisfy a consistency axiom $(\forall P, P' \in V^*)(\forall v \in V)(C_q P = C_q P' = v \Longrightarrow C_q(PP') = v)$ in distributive and some nondistributive lattices, but not, e.g., in lattices of equivalence relations [239, §3.2]. Characterize those lattices where this type of consistency holds.
5.51	Give axiomatic characterizations of *Cen* and *Mea* on the covering graph of $(\mathcal{H}_n, \subseteq)$ for hierarchies on n elements.
6.10	Precisely how strong is S-neutrality? For agreement, consensus, or synthesis rules on phylogenies, characterize those that satisfy *S-Ntr*.
6.11	Does there exist a subtree rule on $\mathcal{H}$ that satisfies *Agr* and *S-Ntr*?
6.12	Semple and Steel [369] have investigated the relationship between the Adams consensus result $C_a P$ for a profile P of hierarchies on the same leaf set and the supertree returned by the MINCUTSUPERTREE algorithm when it is applied to P. They ask if MINCUTSUPERTREE could be modified so that, when applied to hierarchies on the same leaf set, the Adams consensus is equal to the corresponding supertree or can be obtained by contraction from an induced subtree of it [369, p. 157]. If so, could the modified algorithm be characterized axiomatically so that, when restricted to profiles of hierarchies on the same leaf set, it yields a new characterization of the Adams consensus rule?

Bibliography

[1] E. Abouheif and G. A. Wray, *Evolution of the gene network underlying wing polyphenism in ants*, Science, 297 (2002), pp. 249–252.

[2] E. N. Adams III, *Consensus techniques and the comparison of taxonomic trees*, Systematic Zoology, 21 (1972), pp. 390–397.

[3] ——, *n-Trees as nestings: Complexity, similarity, and consensus*, Journal of Classification, 3 (1986), pp. 299–317.

[4] F. T. Aleskerov, *Arrovian Aggregation Models*, no. 39 in Theory and Decision Library Series B: Mathematical and Statistical Methods, Kluwer Academic Publishers, Boston, Massachusetts, 1999.

[5] ——, *Categories of Arrovian voting schemes*, in Arrow et al. [20], ch. 2, pp. 95–129.

[6] A. Amir and D. Keselman, *Maximum agreement subtree in a set of evolutionary trees: Metrics and efficient algorithms*, SIAM Journal on Computing, 26 (1997), pp. 1656–1669.

[7] O. Arkhipoff, *An introduction to the axiomatics of procedures of aggregation*, Mathematical Social Sciences, 1 (1980), pp. 69–83.

[8] T. E. Armstrong, *Arrow's theorem with restricted coalition algebras*, Journal of Mathematical Economics, 7 (1980), pp. 55–75.

[9] G. Arnqvist and D. Wooster, *Meta-analysis: Synthesizing research findings in ecology and evolution*, Trends in Ecology & Evolution, 10 (1995), pp. 236–240.

[10] K. J. Arrow, *A difficulty in the concept of social welfare*, Journal of Political Economy, 58 (1950), pp. 328–346. Reprinted as Arrow [15].

[11] ——, *Social Choice and Individual Values*, no. 12 in Cowles Commission for Research in Economics: Monographs, Wiley, New York, first ed., 1951.

[12] ——, *Le principe de rationalité dans les décisions collectives*, Économie Appliquée, 5 (1952), pp. 469–484. For an English translation see Arrow [17].

[13] ——, *Social Choice and Individual Values*, no. 12 in Cowles Foundation for Research in Economics at Yale University: Monographs, Wiley, New York, second ed., 1963. Reprinted by Yale University Press (New Haven) in 1978.

[14] ——, *Formal theories of social welfare*, in Dictionary of the History of Ideas: Studies of Selected Pivotal Ideas, P. P. Wiener, ed., vol. 4, Charles Scribner, New York, 1973, pp. 276–284. Reprinted as Arrow [16].

[15] ——, *A difficulty in the concept of social choice*, in Social Choice and Justice [18], ch. 1, pp. 1–29. Reprinted from Arrow [10].

[16] ——, *Formal theories of social welfare*, in Social Choice and Justice [18], ch. 9, pp. 115–132. Reprinted from Arrow [14].

[17] ——, *The principle of rationality in collective decisions*, in Social Choice and Justice [18], ch. 3, pp. 45–58. Translated from Arrow [12].

[18] ——, ed., *Social Choice and Justice*, no. 1 in Collected Papers of Kenneth J. Arrow, Belknap Press of Harvard University Press, Cambridge, Massachusetts, 1983.

[19] ——, *Kenneth J. Arrow*, in Breit and Spencer [108], ch. 3, pp. 43–58. Autobiographical lecture presented in 1984.

[20] K. J. Arrow, A. K. Sen, and K. Suzumura, eds., *Handbook of Social Choice and Welfare: Volume* 1, no. 19 in Handbooks in Economics, Elsevier, Amsterdam, 2002.

[21] ——, eds., *Handbook of Social Choice and Welfare: Volume* 2, no. 19 in Handbooks in Economics, Elsevier, Amsterdam, 2004. Forthcoming.

[22] S. P. Avann, *Metric ternary distributive semi-lattices*, Proceedings of the American Mathematical Society, 12 (1961), pp. 407–414.

[23] N. Baigent, *Twitching weak dictators*, Journal of Economics, 47 (1987), pp. 407–411.

[24] ——, *Topological theories of social choice*, in Arrow et al. [21], ch. 17. Forthcoming.

[25] W. Bains, *MULTAN: A program to align multiple DNA sequences*, Nucleic Acids Research, 14 (1986), pp. 159–177.

[26] ——, *MULTAN(2), a multiple string alignment program for nucleic acids and proteins*, Computer Applications in the Biosciences, 5 (1989), pp. 51–52.

[27] H. J. Bandelt and J. P. Barthélemy, *Medians in median graphs*, Discrete Applied Mathematics, 8 (1984), pp. 131–142.

[28] H. J. Bandelt and A. W. M. Dress, *Weak hierarchies associated with similarity measures – An additive clustering technique*, Bulletin of Mathematical Biology, 51 (1989), pp. 133–166.

[29] C. R. M. Bangham and B. Asquith, *Viral immunology from math*, Science, 291 (2001), p. 992.

[30] S. Barberà, *Pivotal voters: A new proof of Arrow's theorem*, Economics Letters, 6 (1980), pp. 13–16.

[31] ——, *Strategy-proofness and pivotal voters: A direct proof of the Gibbard-Satterthwaite theorem*, International Economic Review, 24 (1983), pp. 413–417.

[32] ——, *Strategy proofness*, in Arrow et al. [21], ch. 24. Forthcoming.

[33] M. Barbut, *Quelques aspects mathématiques de la décision rationnelle*, Les Temps Modernes, 15 (1959), pp. 725–745. For an English translation see [34].

[34] ——, *Does the majority ever rule?*, Portfolio and Art News Annual, 4 (1961), pp. 79–83, 161–168. Translation of [33] by Corrine Hoexter.

[35] ——, *Médiane, distributivité, éloignements*, Mathématiques et Sciences humaines, 70 (1980), pp. 5–31. Written in 1961.

[36] ——, *Médianes, Condorcet et Kendall*, Mathématiques et Sciences humaines, 69 (1980), pp. 5–13. Written in 1967.

[37] W. A. Barnett, H. Moulin, M. Salles, and N. J. Schofield, eds., *Social Choice, Welfare, and Ethics: Proceedings of the Eighth International Symposium in Economic Theory and Econometrics*, International Symposia in Economic Theory and Econometrics, Cambridge University Press, Cambridge, 1995.

[38] M. Barrett, M. J. Donoghue, and E. Sober, *Against consensus*, Systematic Zoology, 40 (1991), pp. 486–493.

[39] J. P. Barthélemy, *Sur les éloignements symétriques et le principe de Pareto*, Mathématiques et Sciences humaines, 56 (1976), pp. 97–125.

[40] ——, *Caractérisations axiomatiques de la distance de la différence symétrique entre des relations binaires*, Mathématiques et Sciences humaines, 67 (1979), pp. 85–113.

[41] ——, *Trois propriétés des médianes dans un treillis modulaire*, Mathématiques et Sciences humaines, 75 (1981), pp. 83–91.

[42] ——, *Arrow's theorem: Unusual domains and extended codomains*, Mathematical Social Sciences, 3 (1982), pp. 79–89.

[43] ——, *Comments on "Aggregation of equivalence relations" by P. C. Fishburn and A. Rubinstein*, Journal of Classification, 5 (1988), pp. 85–87.

[44] ——, *Thresholded consensus for n-trees*, Journal of Classification, 5 (1988), pp. 229–236.

[45] ——, *Social welfare and aggregation procedures: Combinatorial and algorithmic aspects*, in Applications of Combinatorics and Graph Theory to the Biological and Social Sciences, F. S. Roberts, ed., no. 17 in IMA Volumes in Mathematics and its Applications, Springer-Verlag, New York, 1989, pp. 39–73.

[46] J. P. Barthélemy and M. F. Janowitz, *A formal theory of consensus*, SIAM Journal on Discrete Mathematics, 4 (1991), pp. 305–322.

[47] J. P. BARTHÉLEMY AND B. LECLERC, *The median procedure for partitions*, in Cox et al. [139], pp. 3–34.

[48] J. P. BARTHÉLEMY, B. LECLERC, AND B. MONJARDET, *On the use of ordered sets in problems of comparison and consensus of classifications*, Journal of Classification, 3 (1986), pp. 187–224.

[49] J. P. BARTHÉLEMY AND F. R. MCMORRIS, *The median procedure for n-trees*, Journal of Classification, 3 (1986), pp. 329–334.

[50] ——, *On an independence condition for consensus n-trees*, Applied Mathematics Letters, 2 (1989), pp. 75–78.

[51] J. P. BARTHÉLEMY, F. R. MCMORRIS, AND R. C. POWERS, *Independence conditions for consensus n-trees revisited*, Applied Mathematics Letters, 4 (1991), pp. 43–46.

[52] ——, *Dictatorial consensus functions on n-trees*, Mathematical Social Sciences, 25 (1992), pp. 59–64.

[53] ——, *Stability conditions for consensus functions defined on n-trees*, Mathematical and Computer Modelling, 22 (1995), pp. 79–87.

[54] J. P. BARTHÉLEMY AND B. MONJARDET, *The median procedure in cluster analysis and social choice theory*, Mathematical Social Sciences, 1 (1981), pp. 235–267.

[55] ——, *The median procedure in data analysis: New results and open problems*, in Classification and Related Methods of Data Analysis: Proceedings of the First Conference of the International Federation of Classification Societies (IFCS), Technical University of Aachen, F.R.G., 29 June–1 July 1987, H. H. Bock, ed., Elsevier, Amsterdam, 1988, pp. 309–316.

[56] Y. M. BARYSHNIKOV, *Unifying impossibility theorems: A topological approach*, Advances in Applied Mathematics, 14 (1993), pp. 404–415.

[57] ——, *Topological and discrete social choice: In search of a theory*, Social Choice and Welfare, 14 (1997), pp. 199–209. Reprinted in Heal [203, pp. 53–63].

[58] A. BATBEDAT, *Les isomorphismes HTS et HTE (apres la bijection de Benzecri/Johnson) (première partie)*, Metron, 46 (1988), pp. 47–59.

[59] ——, *Applications des isomorphismes HTS et HTE (vers la classification arborée) (seconde partie)*, Metron, 47 (1989), pp. 35–51.

[60] P. BATTEAU, J. M. BLIN, AND B. MONJARDET, *Stability of aggregation procedures, ultrafilters, and simple games*, Econometrica, 49 (1981), pp. 527–534.

[61] B. R. BAUM, *Combining trees as a way of combining data sets for phylogenetic inference, and the desirability of combining gene trees*, Taxon, 41 (1992), pp. 3–10.

[62] B. R. BAUM AND M. A. RAGAN, *Reply to A. G. Rodrigo's "A comment on Baum's method for combining phylogenetic trees,"* Taxon, 42 (1993), pp. 637–640.

[63] J. L. BELL AND A. B. SLOMSON, *Models and Ultraproducts: An Introduction*, North-Holland, Amsterdam, 1969.

[64] J. BENNETZEN, *Opening the door to comparative plant biology*, Science, 296 (2002), pp. 60–61, 63.

[65] C. BERGE, *Graphs and Hypergraphs*, no. 6 in North-Holland Mathematical Library, North-Holland, Amsterdam, second revised ed., 1976. Translated by E. Minieka.

[66] J. E. BIAGI, *Facility location and the mean function: A study of centrality measures in facility location with an emphasis on the mean function*, Master's thesis, Department of Mathematics, University of Louisville, Louisville, Kentucky, Oct. 2000.

[67] O. R. P. BININDA-EMONDS, *MRP supertree construction in the consensus setting*, in Janowitz et al. [209], pp. 231–242.

[68] ——, ed., *Phylogenetic Supertrees: Combining Information to Reveal the Tree of Life*, Computational Biology, Kluwer Academic Publishers, Boston, Massachusetts, 2004. Forthcoming.

[69] O. R. P. BININDA-EMONDS AND H. N. BRYANT, *Properties of matrix representation with parsimony analyses*, Systematic Biology, 47 (1998), pp. 497–508.

[70] O. R. P. BININDA-EMONDS, J. L. GITTLEMAN, AND M. A. STEEL, *The (super)tree of life: Procedures, problems, and prospects*, Annual Review of Ecology and Systematics, 33 (2002), pp. 265–289.

[71] O. R. P. BININDA-EMONDS AND M. J. SANDERSON, *Assessment of the accuracy of matrix representation with parsimony analysis supertree construction*, Systematic Biology, 50 (2001), pp. 565–579.

[72] G. BIRKHOFF, *Lattice Theory*, no. 25 in Colloquium Publications, American Mathematical Society, Providence, Rhode Island, third (new) ed., 1967.

[73] D. BLACK, *The decisions of a committee using a special majority*, Econometrica, 16 (1948), pp. 245–261.

[74] ——, *The elasticity of committee decisions with an altering size of majority*, Econometrica, 16 (1948), pp. 262–270.

[75] ——, *On the rationale of group decision-making*, Journal of Political Economy, 56 (1948), pp. 23–34.

[76] ——, *Un approccio alla teoria delle decisioni di comitato*, Giornale degli Economisti e Annali di Economia, 7 (1948 Nuova Serie), pp. 262–284.

[77] ——, *The elasticity of committee decisions with alterations in the members' preference schedules*, South African Journal of Economics, 17 (1949), pp. 88–102.

[78] ——, *Some theoretical schemes of proportional representation*, Canadian Journal of Economics and Political Science, 15 (1949), pp. 334–343.

[79] ——, *The theory of elections in single-member constituencies*, Canadian Journal of Economics and Political Science, 15 (1949), pp. 158–175.

[80] ——, *The unity of political and economic science*, Economic Journal, 60 (1950), pp. 506–514. Reprinted in Black [85, pp. 353–361].

[81] ——, *The Theory of Committees and Elections*, Cambridge University Press, Cambridge, 1958. Reprinted by Kluwer Academic Publishers (Boston) in 1987; also see Black [85].

[82] ——, *On Arrow's impossibility theorem*, Journal of Law and Economics, 12 (1969), pp. 227–248. Reprinted in Black [85, pp. 369–385].

[83] ——, *Partial justification of the Borda count*, Public Choice, 28 (1976), pp. 1–15. Reprinted in Black [85, pp. 331–352].

[84] ——, *Arrow's work and the normative theory of committees*, Journal of Theoretical Politics, 3 (1991), pp. 259–275. I. McLean and D. Squires, eds. Written in March 1972; published posthumously; reprinted in Black [85, pp. 387–405].

[85] ——, *The Theory of Committees and Elections, by Duncan Black, and Committee Decisions with Complementary Valuation, by Duncan Black and R. A. Newing*, Kluwer Academic Publishers, Boston, Massachusetts, revised second ed., 1998. I. McLean, A. McMillan and B. L. Monroe, eds.

[86] D. Black and R. A. Newing, *Committee Decisions with Complementary Valuation*, William Hodge, Edinburgh, 1951. Written in 1949; reprinted in Black [85, pp. 273–327].

[87] D. H. Blair, G. Bordes, J. S. Kelly, and K. Suzumura, *Impossibility theorems without collective rationality*, Journal of Economic Theory, 13 (1976), pp. 361–379.

[88] D. H. Blair and R. A. Pollak, *Collective rationality and dictatorship: The scope of the Arrow theorem*, Journal of Economic Theory, 21 (1979), pp. 186–194.

[89] J. H. Blau, *The existence of social welfare functions*, Econometrica, 25 (1957), pp. 302–313.

[90] ——, *Social choice functions and simple games*, Bulletin of the American Mathematical Society, 63 (1957), pp. 243–244.

[91] ——, *Arrow's theorem with weak independence*, Economica, 38 (1971), pp. 413–420.

[92] ——, *A direct proof of Arrow's theorem*, Econometrica, 40 (1972), pp. 61–67.

[93] ——, *Semiorders and collective choice*, Journal of Economic Theory, 21 (1979), pp. 195–206.

[94] S. D. Bloomfield, *A social choice interpretation of the von Neumann-Morgenstern game*, Econometrica, 44 (1976), pp. 105–114.

[95] S. D. BLOOMFIELD AND R. B. WILSON, *The postulates of game theory*, Journal of Mathematical Sociology, 2 (1972), pp. 221–234.

[96] S. BÖCKER, D. BRYANT, A. W. M. DRESS, AND M. A. STEEL, *Algorithmic aspects of tree amalgamation*, Journal of Algorithms, 37 (2000), pp. 522–537.

[97] S. BÖCKER, A. W. M. DRESS, AND M. A. STEEL, *Patching up X-trees*, Annals of Combinatorics, 3 (1999), pp. 1–12.

[98] K. P. BOGART, *Preference structures* I*: Distances between transitive preference relations*, Journal of Mathematical Sociology, 3 (1973), pp. 49–67.

[99] ——, *Preference structures* II*: Distances between asymmetric relations*, SIAM Journal on Applied Mathematics, 29 (1975), pp. 254–262.

[100] J. C. D. BORDA, *Mémoire sur les Élections au Scrutin*, in Mémoires de l'Académie Royale des Sciences année 1781, 1784, pp. 657–665. Read to the Academy in 1784, then published that year in the Mémoires for 1781. For an English translation see de Grazia [186] or McLean and Urken [258, chap. 5].

[101] G. BORDES, *Consistency, rationality and collective choice*, Review of Economic Studies, 43 (1976), pp. 451–457.

[102] G. BORDES AND N. TIDEMAN, *Independence of irrelevant alternatives in the theory of voting*, Theory and Decision, 30 (1991), pp. 163–186.

[103] N. BOURBAKI, *General Topology: Part* 1, no. 3 in Elements of Mathematics, Addison-Wesley, Reading, Massachusetts, 1966.

[104] ——, *Theory of Sets*, no. 1 in Elements of Mathematics, Addison-Wesley, Reading, Massachusetts, 1968.

[105] D. BOUYSSOU, *Democracy and efficiency: A note on "Arrow's theorem is not a surprising result"*, European Journal of Operational Research, 58 (1992), pp. 427–430.

[106] G. L. BRADY AND G. TULLOCK, eds., *Formal Contributions to the Theory of Public Choice: The Unpublished Works of Duncan Black*, Kluwer Academic Publishers, Boston, Massachusetts, 1996.

[107] S. J. BRAMS AND P. C. FISHBURN, *Voting procedures*, in Arrow et al. [20], ch. 4, pp. 173–236.

[108] W. BREIT AND R. W. SPENCER, eds., *Lives of the Laureates: Thirteen Nobel Economists*, MIT Press, Cambridge, Massachusetts, third ed., 1995.

[109] K. BREMER, *Combinable component consensus*, Cladistics: The International Journal of the Willi Hennig Society, 6 (1990), pp. 369–372.

[110] D. J. BROWN, *An approximate solution to Arrow's problem*, Journal of Economic Theory, 9 (1974), pp. 375–383.

[111] ——, *Aggregation of preferences*, Quarterly Journal of Economics, 89 (1975), pp. 456–469.

[112] D. Bryant, *Building Trees, Hunting for Trees, and Comparing Trees: Theory and Methods in Phylogenetic Analysis*, PhD thesis, University of Canterbury, Department of Mathematics, 1997. xiv+212 pp.

[113] ——, *A classification of consensus methods for phylogenies*, in Janowitz et al. [209], pp. 163–183.

[114] D. Bryant, A. McKenzie, and M. A. Steel, *The size of a maximum agreement subtree for random binary trees*, in Janowitz et al. [209], pp. 55–65.

[115] D. E. Campbell, *Democratic preference functions*, Journal of Economic Theory, 12 (1976), pp. 259–272.

[116] ——, *Algorithms for social choice functions*, Review of Economic Studies, 47 (1980), pp. 617–627.

[117] ——, *On the derivation of majority rule*, Theory and Decision, 14 (1982), pp. 133–140.

[118] ——, *A characterization of simple majority rule for restricted domains*, Economics Letters, 28 (1988), pp. 307–310.

[119] ——, *Equity, Efficiency, and Social Choice*, Clarendon Press, Oxford, 1992.

[120] D. E. Campbell and J. S. Kelly, *t or $1 - t$. That is the trade-off*, Econometrica, 61 (1993), pp. 1355–1365.

[121] ——, *Nondictatorially independent pairs*, Social Choice and Welfare, 12 (1995), pp. 75–86.

[122] ——, *A simple characterization of majority rule*, Economic Theory, 15 (2000), pp. 689–700.

[123] ——, *Weak independence and veto power*, Economics Letters, 66 (2000), pp. 183–189.

[124] ——, *Impossibility theorems in the Arrovian framework*, in Arrow et al. [20], ch. 1, pp. 35–94.

[125] ——, *On the Arrow and Wilson impossibility theorems*, Social Choice and Welfare, 20 (2003), pp. 273–281.

[126] D. Chen, L. Diao, O. Eulenstein, D. Fernández-Baca, and M. Sanderson, *Flipping: A supertree construction method*, in Janowitz et al. [209], pp. 135–160.

[127] V. Chepoi and B. Fichet, *l_∞-Approximation via subdominants*, Journal of Mathematical Psychology, 44 (2000), pp. 600–616.

[128] G. Chichilnisky, *Social choice and the topology of spaces of preferences*, Advances in Mathematics, 37 (1980), pp. 165–176.

[129] ——, *Social aggregation rules and continuity*, Quarterly Journal of Economics, 97 (1982), pp. 337–352.

[130] ——, *Structural instability of decisive majority rules*, Journal of Mathematical Economics, 9 (1982), pp. 207–221.

[131] ——, *The topological equivalence of the Pareto condition and the existence of a dictator*, Journal of Mathematical Economics, 9 (1982), pp. 223–233.

[132] G. Chichilnisky and G. M. Heal, *Necessary and sufficient conditions for a resolution of the social choice paradox*, Journal of Economic Theory, 31 (1983), pp. 68–87.

[133] ——, *Social choice with infinite populations: Construction of a rule and impossibility results*, Social Choice and Welfare, 14 (1997), pp. 303–318. Written in 1979. Reprinted in Heal [203, pp. 157–172].

[134] J. E. Cohen, *Can fitness be aggregated?*, American Naturalist, 125 (1985), pp. 716–729.

[135] R. Cole, M. Farach-Colton, R. Hariharan, T. Przytycka, and M. Thorup, *An $O(n \log n)$ algorithm for the maximum agreement subtree problem for binary trees*, SIAM Journal on Computing, 30 (2000), pp. 1385–1404.

[136] H. Colonius and H. H. Schulze, *Tree structures for proximity data*, British Journal of Mathematical & Statistical Psychology, 34 (1981), pp. 167–180.

[137] M. d. Condorcet, M. J. A. N. Caritat, *Essai sur l'Application de l'Analyse à la Probabilité des Décisions Rendues à la Pluralité des Voix*, De l'Imprimerie royale, Paris, 1785. Photographic reprint in 1972 by Chelsea Publishing Co. (New York), cxci+304 pp.

[138] M. Constantinescu and D. Sankoff, *An efficient algorithm for supertrees*, Journal of Classification, 12 (1995), pp. 101–112.

[139] I. J. Cox, P. Hansen, and B. Julesz, eds., *Partitioning Data Sets: DIMACS Workshop, April 19-21, 1993*, no. 19 in DIMACS Series in Discrete Mathematics and Theoretical Computer Science, American Mathematical Society, Providence, Rhode Island, 1995.

[140] R. A. Cramer-Benjamin, *Independence in the Ordinal Model and on Closed Set Systems*, PhD thesis, University of Massachusetts, Amherst, Department of Mathematics and Statistics, Sept. 1998. ix+69 pp.

[141] R. A. Cramer-Benjamin, G. D. Crown, and M. F. Janowitz, *Permutation invariance*, Algebra Universalis, (2003). Forthcoming.

[142] P. Crawley and R. P. Dilworth, *Algebraic Theory of Lattices*, Prentice-Hall, Englewood Cliffs, New Jersey, 1973.

[143] G. D. Crown and M. F. Janowitz, *An injective set representation of closed systems of sets*, in Janowitz et al. [209], pp. 67–79.

[144] G. D. Crown, M. F. Janowitz, and R. C. Powers, *Neutral consensus functions*, Mathematical Social Sciences, 25 (1993), pp. 231–250.

[145] ———, *An ordered set approach to neutral consensus functions*, in Diday et al. [155], pp. 102–110.

[146] ———, *Further results on neutral consensus functions*, Mathématiques, Informatique et Sciences humaines, 132 (1995), pp. 5–11.

[147] M. S. Daskin, *Network and Discrete Location: Models, Algorithms, and Applications*, Wiley-Interscience Series in Discrete Mathematics and Optimization, Wiley, New York, 1995.

[148] B. A. Davey and H. A. Priestley, *Introduction to Lattices and Order*, Cambridge University Press, Cambridge, second ed., 2002.

[149] W. H. E. Day, *Optimal algorithms for comparing trees with labeled leaves*, Journal of Classification, 2 (1985), pp. 7–28.

[150] ———, ed., *Special issue: Consensus classifications*, Journal of Classification, 3 (1986), pp. 183–356. Issue no. 2.

[151] W. H. E. Day and F. R. McMorris, *Analysing molecular sequences using consensus*, New Zealand Journal of Botany, 31 (1993), pp. 211–218.

[152] ———, *The computation of consensus patterns in DNA sequences*, Mathematical and Computer Modelling, 17 (1993), pp. 49–52.

[153] ———, *Axiomatics in group choice and bioconsensus*, in Janowitz et al. [209], pp. 3–35.

[154] W. H. E. Day, F. R. McMorris, and D. B. Meronk, *Axioms for consensus functions based on lower bounds in posets*, Mathematical Social Sciences, 12 (1986), pp. 185–190.

[155] E. Diday, Y. Lechevallier, M. Schader, P. Bertrand, and B. Burtschy, eds., *New Approaches in Classification and Data Analysis*, Studies in Classification, Data Analysis, and Knowledge Organization, Springer-Verlag, Berlin, 1994.

[156] F. Domenach and B. Leclerc, *Closure systems, implicational systems, overhanging relations and the case of hierarchical classification*, Mathematical Social Sciences, (2003). Forthcoming.

[157] M. Dwyer, F. R. McMorris, and R. C. Powers, *Removal independence and multi-consensus functions*, Mathématiques, Informatique et Sciences humaines, 148 (1999), pp. 31–40.

[158] G. F. Estabrook, *Ancestor-descendant relations and incompatible data: Motivation for research in discrete mathematics*, in Mirkin et al. [286], pp. 1–28.

[159] M. Farach, T. M. Przytycka, and M. Thorup, *On the agreement of many trees*, Information Processing Letters, 55 (1995), pp. 297–301.

[160] J. Feldman Högaasen, *Ordres partiels et permutoèdre*, Mathématiques et Sciences humaines, 28 (1969), pp. 27–38.

[161] J. Felsenstein, *Inferring Phylogenies*, Sinauer Associates, Sunderland, Massachusetts, 2003. Forthcoming.

[162] C. R. Finden and A. D. Gordon, *Obtaining common pruned trees*, Journal of Classification, 2 (1985), pp. 255–276.

[163] P. C. Fishburn, *Arrow's impossibility theorem: Concise proof and infinite voters*, Journal of Economic Theory, 2 (1970), pp. 103–106.

[164] ——, *Summation social choice functions*, Econometrica, 41 (1973), pp. 1183–1196.

[165] ——, *On collective rationality and a generalized impossibility theorem*, Review of Economic Studies, 41 (1974), pp. 445–457.

[166] ——, *Interprofile Conditions and Impossibility*, no. 18 in Fundamentals of pure and applied economics, Harwood Academic Publishers, Chur, Switzerland, 1987.

[167] P. C. Fishburn and W. V. Gehrlein, *Borda's rule, positional voting, and Condorcet's simple majority principle*, Public Choice, 28 (1976), pp. 79–88.

[168] P. C. Fishburn and A. Rubinstein, *Aggregation of equivalence relations*, Journal of Classification, 3 (1986), pp. 61–65. But see Barthélemy [43].

[169] T. Flanagan, *The Tenants of Time*, Warner Books, New York, 1988.

[170] W. Gaertner, *Domain restrictions*, in Arrow et al. [20], ch. 3, pp. 131–170.

[171] D. J. Galas, M. Eggert, and M. S. Waterman, *Rigorous pattern-recognition methods for DNA sequences: Analysis of promoter sequences from Escherichia coli*, Journal of Molecular Biology, 186 (1985), pp. 117–128.

[172] P. Gärdenfors, *Positionalist voting functions*, Theory and Decision, 4 (1973), pp. 1–24.

[173] ——, *Manipulation of social choice functions*, Journal of Economic Theory, 13 (1976), pp. 217–228.

[174] M. R. Garey and D. S. Johnson, *Computers and Intractability: A Guide to the Theory of NP-Completeness*, Series of Books in the Mathematical Sciences, W. H. Freeman, San Francisco, 1979.

[175] J. GEANAKOPLOS, *Three brief proofs of Arrow's impossibility theorem*, Cowles Foundation Discussion Paper 1123RRR, Cowles Foundation for Research in Economics, Yale University, Box 208281, New Haven, Connecticut 06520-8281, June 2001. Obtained from http://cowles.econ.yale.edu/.

[176] W. V. GEHRLEIN, *Condorcet's paradox and the Condorcet efficiency of voting rules*, Mathematica Japonica, 45 (1997), pp. 173–199.

[177] W. V. GEHRLEIN AND D. LEPELLEY, *The Condorcet efficiency of Borda rule with anonymous voters*, Mathematical Social Sciences, 41 (2001), pp. 39–50.

[178] A. GIBBARD, *Manipulation of voting schemes: A general result*, Econometrica, 41 (1973), pp. 587–601.

[179] W. D. GODDARD, E. KUBICKA, G. KUBICKI, AND F. R. MCMORRIS, *The agreement metric for labeled binary trees*, Mathematical Biosciences, 123 (1994), pp. 215–226.

[180] ——, *Agreement subtrees, metric and consensus for labeled binary trees*, in Cox et al. [139], pp. 97–104.

[181] W. D. GODDARD AND G. KUBICKI, *The minimum size of agreement subtrees of two binary trees*, Congressus Numerantium, 97 (1993), pp. 131–136.

[182] A. D. GORDON, *On the assessment and comparison of classifications*, in Analyse de Données et Informatique: Fontainebleau, du 19 au 30 mars 1979: Cours de la Commission des Communautes Europeennes, R. Tomassone, ed., Institut National de Recherche en Informatique et en Automatique, Le Chesnay, 1980, pp. 149–160.

[183] ——, *Consensus supertrees: The synthesis of rooted trees containing overlapping sets of labeled leaves*, Journal of Classification, 3 (1986), pp. 335–348.

[184] ——, *Hierarchical classification*, in Clustering and Classification, P. Arabie, L. J. Hubert, and G. D. Soete, eds., World Scientific, Singapore, 1996, pp. 65–121.

[185] G. GRÄTZER, *General Lattice Theory*, no. 75 in Pure and Applied Mathematics: A Series of Monographs and Textbooks, Academic Press, New York, 1978.

[186] A. D. GRAZIA, *Mathematical derivation of an election system*, Isis, 44 (1953), pp. 42–51. A translation into English of Borda [100].

[187] J. GREENBERG, *Consistent majority rules over compact sets of alternatives*, Econometrica, 47 (1979), pp. 627–636.

[188] B. GROFMAN AND S. L. FELD, *Rousseau's General Will: A Condorcetian perspective*, American Political Science Review, 82 (1988), pp. 567–576.

[189] A. S. GUHA, *Neutrality, monotonicity, and the right of veto*, Econometrica, 40 (1972), pp. 821–826.

[190] G. T. GUILBAUD, *Les théories de l'intérêt général et le problème logique de l'agrégation*, Économie Appliquée, 5 (1952), pp. 501–584. For an English translation see [191].

[191] ——, *Theories of the general interest, and the logical problem of aggregation*, in Readings in Mathematical Social Science, P. F. Lazarsfeld and N. W. Henry, eds., Science Research Associates, Chicago, 1966, pp. 262–307. Reprinted by MIT Press (Cambridge, Massachusetts) in 1968.

[192] G. T. GUILBAUD AND P. ROSENSTIEHL, *Analyse algébrique d'un scrutin*, Mathématiques et Sciences humaines, 4 (1963), pp. 9–33. Reprinted (and slightly expanded) in [193].

[193] ——, *Analyse algébrique d'un scrutin*, in Ordres Totaux Finis: Travaus du Séminaire sur les Ordres Totaux Finis, Aix-en-Provence, Juillet 1967, C. de Mathématique Sociale de l'École Pratique des Hautes Etudes, ed., no. 12 in Mathématiques et Sciences de l'Homme, Mouton, Paris, 1971, pp. 71–100. A slightly expanded version of [192].

[194] A. GUPTA AND N. NISHIMURA, *Finding largest subtrees and smallest supertrees*, Algorithmica: An International Journal in Computer Science, 21 (1998), pp. 183–210.

[195] D. GUSFIELD, *Algorithms on Strings, Trees, and Sequences: Computer Science and Computational Biology*, Cambridge University Press, Cambridge, 1997. Reprinted 1999 with corrections.

[196] G. Y. HANDLER AND P. B. MIRCHANDANI, *Location on Networks: Theory and Algorithms*, MIT Press Series in Signal Processing, Optimization, and Control, MIT Press, Cambridge, Massachusetts, 1979.

[197] B. HANSSON, *Group preferences*, Econometrica, 37 (1969), pp. 50–54.

[198] ——, *Voting and group decision functions*, Synthese, 20 (1969), pp. 526–537.

[199] ——, *The independence condition in the theory of social choice*, Theory and Decision, 4 (1973), pp. 25–49.

[200] ——, *The existence of group preference functions*, Public Choice, 28 (1976), pp. 89–98. Written in late 1971 and first published as Working Paper No. 3 in the mimeographed series of the Mattias Fremling Society.

[201] S. O. HANSSON, *A procedural model of voting*, Theory and Decision, 32 (1992), pp. 269–301.

[202] F. HARARY, *Graph Theory*, Addison-Wesley, Reading, Massachusetts, 1969.

[203] G. M. HEAL, ed., *Topological Social Choice*, Springer-Verlag, Berlin, 1997. First published as Special Issue (Volume 14, Number 2, 1997) in *Social Choice and Welfare*, and so includes [57, 277, 359, 428].

[204] M. R. HENZINGER, V. KING, AND T. WARNOW, *Constructing a tree from homeomorphic subtrees, with applications to computational evolutionary biology*, Algorithmica: An International Journal in Computer Science, 24 (1999), pp. 1–13.

[205] D. M. HILLIS, *Molecular versus morphological approaches to systematics*, Annual Review of Ecology and Systematics, 18 (1987), pp. 23–42.

[206] R. HOLZMAN, *An axiomatic approach to location on networks*, Mathematics of Operations Research, 15 (1990), pp. 553–563.

[207] E. V. HUNTINGTON, *A paradox in the scoring of competing teams*, Science, 88 (1938), pp. 287–288. Postulate of relevancy = independence.

[208] S. ISHIKAWA AND K. NAKAMURA, *The strategy-proof social choice functions*, Journal of Mathematical Economics, 6 (1979), pp. 283–295.

[209] M. F. JANOWITZ, F. J. LAPOINTE, F. R. MCMORRIS, B. G. MIRKIN, AND F. S. ROBERTS, eds., *Bioconsensus*, no. 61 in DIMACS Series in Discrete Mathematics and Theoretical Computer Science, American Mathematical Society, Providence, Rhode Island, 2003.

[210] N. JARDINE AND R. SIBSON, *The construction of hierarchic and non-hierarchic classifications*, Computer Journal, 11 (1968), pp. 177–184.

[211] ——, *A model for taxonomy*, Mathematical Biosciences, 2 (1968), pp. 465–482.

[212] ——, *Mathematical Taxonomy*, Wiley Series in Probability and Mathematical Statistics, Wiley, London, 1971.

[213] S. C. JOHNSON, *Hierarchical clustering schemes*, Psychometrika, 32 (1967), pp. 241–254.

[214] B. JONES, B. RADCLIFF, C. TABER, AND R. TIMPONE, *Condorcet winners and the paradox of voting: Probability calculations for weak preference orders*, American Political Science Review, 89 (1995), pp. 137–144.

[215] G. KALAI, *A Fourier-theoretic perspective on the Condorcet paradox and Arrow's theorem*, Advances in Applied Mathematics, 29 (2002), pp. 412–426.

[216] S. KANNAN, T. WARNOW, AND S. YOOSEPH, *Computing the local consensus of trees*, SIAM Journal on Computing, 27 (1998), pp. 1695–1724.

[217] M. Y. KAO, *Tree contractions and evolutionary trees*, SIAM Journal on Computing, 27 (1998), pp. 1592–1616.

[218] I. KAPLANSKY, *Set Theory and Metric Spaces*, Allyn and Bacon Series in Advanced Mathematics, Chelsea, New York, second ed., 1977.

[219] J. S. KELLY, *The continuous representation of a social preference ordering*, Econometrica, 39 (1971), pp. 593–597.

[220] ——, *Arrow Impossibility Theorems*, Economic Theory and Mathematical Economics, Academic Press, New York, 1978.

[221] ——, *The free triple assumption*, Social Choice and Welfare, 11 (1994), pp. 97–101.

[222] J. G. KEMENY, *Mathematics without numbers*, Daedalus: Proceedings of the American Academy of Arts and Sciences, 88 (1959), pp. 577–591.

[223] J. G. KEMENY AND J. L. SNELL, *Mathematical Models in the Social Sciences*, Introductions to Higher Mathematics, Ginn, Boston, 1962. Reprinted by MIT Press (Cambridge, Massachusetts) in 1972 and 1975.

[224] ——, *Preference rankings: An axiomatic approach*, in Mathematical Models in the Social Sciences [223], ch. 2, pp. 9–23. Reprinted by MIT Press (Cambridge, Massachusetts) in 1972 and 1975.

[225] D. KESELMAN AND A. AMIR, *Maximum agreement subtree in a set of evolutionary trees—metrics and efficient algorithms*, in 35th Annual Symposium on Foundations of Computer Science: Proceedings, S. Goldwasser, ed., IEEE Computer Society Press, Los Alamitos, California, 1994, pp. 758–769.

[226] A. P. KIRMAN AND D. SONDERMANN, *Arrow's theorem, many agents, and invisible dictators*, Journal of Economic Theory, 5 (1972), pp. 267–277.

[227] E. KUBICKA, G. KUBICKI, AND F. R. MCMORRIS, *On agreement subtrees of two binary trees*, Congressus Numerantium, 88 (1992), pp. 217–224.

[228] ——, *An algorithm to find agreement subtrees*, Journal of Classification, 12 (1995), pp. 91–99.

[229] ——, *Agreement metrics for trees revisited*, in Mirkin et al. [286], pp. 239–248.

[230] S. M. LANYON, *Phylogenetic frameworks: Towards a firmer foundation for the comparative approach*, Biological Journal of the Linnean Society, 49 (1993), pp. 45–61.

[231] F. J. LAPOINTE AND G. CUCUMEL, *The average consensus procedure: Combination of weighted trees containing identical or overlapping sets of taxa*, Systematic Biology, 46 (1997), pp. 306–312.

[232] L. LAUWERS, *Topological social choice*, Mathematical Social Sciences, 40 (2000), pp. 1–39.

[233] B. LECLERC, *Efficient and binary consensus functions on transitively valued relations*, Mathematical Social Sciences, 8 (1984), pp. 45–61.

[234] ——, *Medians and majorities in semimodular lattices*, SIAM Journal on Discrete Mathematics, 3 (1990), pp. 266–276.

[235] ——, *Aggregation of fuzzy preferences: A theoretic Arrow-like approach*, Fuzzy Sets and Systems, 43 (1991), pp. 291–309.

[236] ——, *Lattice valuations, medians and majorities*, Discrete Mathematics, 111 (1993), pp. 345–356.

[237] ——, *Medians for weight metrics in the covering graphs of semilattices*, Discrete Applied Mathematics, 49 (1994), pp. 281–297.

[238] ——, *Consensus of classifications: The case of trees*, in Advances in Data Science and Classification: Proceedings of the 6th Conference of the International Federation of Classification Societies (IFCS-98), Università "La Sapienza," Rome, 21-24 July, 1998, A. Rizzi, M. Vichi, and H. H. Bock, eds., Studies in Classification, Data Analysis, and Knowledge Organization, Springer-Verlag, Berlin, 1998, pp. 81–90.

[239] ——, *The median procedure in the semilattice of orders*, Discrete Applied Mathematics, 127 (2003), pp. 285–302.

[240] B. Leclerc and B. Monjardet, *Latticial theory of consensus*, in Barnett et al. [37], ch. 6, pp. 145–160.

[241] J. Lehel, F. R. McMorris, and R. C. Powers, *Consensus methods for pyramids and other hypergraphs*, in Data Science, Classification, and Related Methods: Proceedings of the Fifth Conference of the International Federation of Classification Societies (IFCS-96), Kobe, Japan, March 27–30, 1996, C. Hayashi, N. Ohsumi, K. Yajima, Y. Tanaka, H. H. Bock, and Y. Baba, eds., Studies in Classification, Data Analysis, and Knowledge Organization, Springer-Verlag, Tokyo, 1998, pp. 187–190.

[242] M. Lilla, *The new age of tyranny*, New York Review of Books, 49 (2002), pp. 28–29. No. 16, 24 October 2002.

[243] M. Malawski and L. Zhou, *A note on social choice theory without the Pareto principle*, Social Choice and Welfare, 11 (1994), pp. 103–107.

[244] T. Marchant, *Valued relations aggregation with the Borda method*, Journal of Multi-Criteria Decision Analysis, 5 (1996), pp. 127–132.

[245] T. Margush, *Distances between trees*, Discrete Applied Mathematics, 4 (1982), pp. 281–290.

[246] T. Margush and F. R. McMorris, *Consensus n-trees*, Bulletin of Mathematical Biology, 43 (1981), pp. 239–244.

[247] A. Mas-Colell and H. F. Sonnenschein, *General possibility theorems for group decisions*, Review of Economic Studies, 39 (1972), pp. 185–192.

[248] E. S. Maskin, *Majority rule, social welfare functions, and game forms*, in Choice, Welfare, and Development: A Festschrift in Honour of Amartya K. Sen, K. Basu, P. K. Pattanaik, and K. Suzumura, eds., Clarendon Press, Oxford, 1995, ch. 6, pp. 100–109.

[249] K. O. May, *A set of independent necessary and sufficient conditions for simple majority decision*, Econometrica, 20 (1952), pp. 680–684.

[250] ——, *A note on the complete independence of the conditions for simple majority decision*, Econometrica, 21 (1953), pp. 172–173.

[251] J. B. McGuire, *On the consensus construction of an evolutionary tree*, Journal of Social and Biological Structures, 2 (1979), pp. 107–118.

[252] J. B. McGuire and C. J. Thompson, *On the reconstruction of an evolutionary order*, Journal of Theoretical Biology, 75 (1978), pp. 141–147.

[253] I. McLean, *The first golden age of social choice, 1784–1803*, in Barnett et al. [37], ch. 1, pp. 13–33.

[254] ——, *Independence of irrelevant alternatives before Arrow*, Mathematical Social Sciences, 30 (1995), pp. 107–126.

[255] ——, *E. J. Nanson, social choice, and electoral reform*, Australian Journal of Political Science, 31 (1996), pp. 369–385.

[256] I. McLean and F. Hewitt, eds., *Condorcet: Foundations of Social Choice and Political Theory*, Edward Elgar, Aldershot, Hants, England, 1994.

[257] I. McLean and J. London, *The Borda and Condorcet principles: Three medieval applications*, Social Choice and Welfare, 7 (1990), pp. 99–108.

[258] I. McLean and A. B. Urken, eds., *Classics of Social Choice*, University of Michigan Press, Ann Arbor, 1995.

[259] F. R. McMorris, *Axioms for consensus functions on undirected phylogenetic trees*, Mathematical Biosciences, 74 (1985), pp. 17–21.

[260] ——, *The median procedure for n-trees as a maximum likelihood method*, Journal of Classification, 7 (1990), pp. 77–80.

[261] ——, *The median function on structured metric spaces*, Student, 2 (1997), pp. 195–201.

[262] ——, *A view of some centrality and consensus functions in classification theory and beyond*, in Developments in Statistics, A. Mrvar and A. Ferligoj, eds., Faculty of Social Sciences, University of Ljubljana, Slovenia, 2002, pp. 21–29.

[263] F. R. McMorris, D. B. Meronk, and D. A. Neumann, *A view of some consensus methods for trees*, in Numerical Taxonomy, J. Felsenstein, ed., no. 1 in NATO Advanced Science Institutes Series G: Ecological Sciences, Springer-Verlag, Berlin, 1983, pp. 122–126.

[264] F. R. McMorris, H. M. Mulder, and R. C. Powers, *The median function on median graphs and semilattices*, Discrete Applied Mathematics, 101 (2000), pp. 221–230.

[265] ——, *The median function on distributive semilattices*, Discrete Applied Mathematics, 127 (2003), pp. 319–324.

[266] F. R. McMorris, H. M. Mulder, and F. S. Roberts, *The median procedure on median graphs*, Discrete Applied Mathematics, 84 (1998), pp. 165–181.

[267] F. R. McMorris and D. A. Neumann, *Consensus functions defined on trees*, Mathematical Social Sciences, 4 (1983), pp. 131–136.

[268] F. R. McMorris and R. C. Powers, *Consensus weak hierarchies*, Bulletin of Mathematical Biology, 53 (1991), pp. 679–684.

[269] ——, *Consensus functions on trees that satisfy an independence axiom*, Discrete Applied Mathematics, 47 (1993), pp. 47–55.

[270] ——, *The median procedure in a formal theory of consensus*, SIAM Journal on Discrete Mathematics, 8 (1995), pp. 507–516.

[271] ——, *Intersection rules for consensus hierarchies and pyramids*, in Ordinal and Symbolic Data Analysis: Proceedings of the International Conference on Ordinal and Symbolic Data Analysis - OSDA 95, Paris, June 20–23, 1995, E. Diday, Y. Lechevallier, and O. Opitz, eds., Studies in Classification, Data Analysis, and Knowledge Organization, Springer-Verlag, Berlin, 1996, pp. 301–308.

[272] ——, *The median function on weak hierarchies*, in Mirkin et al. [286], pp. 265–269.

[273] ——, *The Arrovian program from weak orders to hierarchial and tree-like relations*, in Janowitz et al. [209], pp. 37–45.

[274] F. R. McMorris, F. S. Roberts, and C. Wang, *The center function on trees*, Networks, 38 (2001), pp. 84–87.

[275] F. R. McMorris and M. A. Steel, *The complexity of the median procedure for binary trees*, in Diday et al. [155], pp. 136–140.

[276] C. A. Meacham, *The loose consensus of evolutionary histories*, in 16th Numerical Taxonomy Conference, Notre Dame, Ind., 1982. Abstract.

[277] P. Mehta, *Topological methods in social choice: An overview*, Social Choice and Welfare, 14 (1997), pp. 233–243. Reprinted in Heal [203, pp. 87–97].

[278] P. Michaud, *Hommage à Condorcet: Version intégrale pour le bicentenaire de l'Essai de Condorcet*, Tech. Rep. F.094, Centre Scientifique IBM, Paris, 1985.

[279] M. F. Mickevich, *Taxonomic Congruence*, PhD thesis, State University of New York at Stony Brook, Department of Ecology and Evolution, 1978. viii+70 pp.

[280] ——, *Taxonomic congruence*, Systematic Zoology, 27 (1978), pp. 143–158.

[281] P. B. Mirchandani and R. L. Francis, eds., *Discrete Location Theory*, Wiley-Interscience Series in Discrete Mathematics and Optimization, Wiley, New York, 1990.

[282] B. G. Mirkin, *On the problem of reconciling partitions*, in Quantitative Sociology: International Perspectives on Mathematical and Statistical Modelling, H. M. Blalock, A. Aganbegian, F. M. Borodkin, R. Boudon, and V. Capecchi, eds., Quantitative Studies in Social Relations, Academic Press, New York, 1975, ch. 15, pp. 441–449.

[283] ——, *Axiomatic analysis of the problems of consistency of relations*, in Group Choice [284], ch. 3, pp. 117–139. Translated from Russian by Y. Oliker.

[284] ——, *Group Choice*, Scripta Series in Mathematics, V. H. Winston, Washington, District of Columbia, 1979. Translated from Russian by Y. Oliker.

[285] ——, *Federations and transitive group choice*, Mathematical Social Sciences, 2 (1982), pp. 35–38.

[286] B. G. Mirkin, F. R. McMorris, F. S. Roberts, and A. Rzhetsky, eds., *Mathematical Hierarchies and Biology: DIMACS Workshop, November 13–15, 1996*, no. 37 in DIMACS Series in Discrete Mathematics and Theoretical Computer Science, American Mathematical Society, Providence, Rhode Island, 1997.

[287] B. G. Mirkin and F. S. Roberts, *Consensus functions and patterns in molecular sequences*, Bulletin of Mathematical Biology, 55 (1993), pp. 695–713.

[288] M. M. Miyamoto, *Consensus cladograms and general classifications*, Cladistics: The International Journal of the Willi Hennig Society, 1 (1985), pp. 186–189.

[289] B. Monjardet, *Tournois et ordres médians pour une opinion*, Mathématiques et Sciences humaines, 43 (1973), pp. 55–70.

[290] ——, *An axiomatic theory of tournament aggregation*, Mathematics of Operations Research, 3 (1978), pp. 334–351.

[291] ——, *Duality in the theory of social choice*, in Aggregation and Revelation of Preferences, J. J. Laffont, ed., no. 2 in Studies in Public Economics, North-Holland, Amsterdam, 1979, ch. 7, pp. 131–143.

[292] ——, *Théorie de la médiane dans les treillis distributifs finis et applications*, Annals of Discrete Mathematics, 9 (1980), pp. 87–91.

[293] ——, *Variations sur l'effet Condorcet ou Condorcet, Black, Arrow, Guilbaud ...et les autres*, Cahiers du Centre d'Études de Recherche Opérationnelle, 22 (1980), pp. 7–15.

[294] ——, *Metrics on partially ordered sets – A survey*, Discrete Mathematics, 35 (1981), pp. 173–184.

[295] ——, *On the use of ultrafilters in social choice theory*, in Social Choice and Welfare, P. K. Pattanaik and M. Salles, eds., no. 145 in Contributions to Economic Analysis, North-Holland, Amsterdam, 1983, ch. 5, pp. 73–78.

[296] ——, *Arrowian characterizations of latticial federation consensus functions*, Mathematical Social Sciences, 20 (1990), pp. 51–71.

[297] ——, *Sur diverses formes de la 'Règle de Condorcet' d'agrégation des préférences*, Mathématiques, Informatique et Sciences humaines, 111 (1990), pp. 61–71.

[298] ——, *Éléments pour une histoire de la médiane métrique*, in Moyenne, Milieu, Centre: Histoires et Usages, J. Feldman, G. Lagneau, and B. Matalon, eds., no. 5 in Histoire des Sciences et des Techniques, Éditions de l'École des Hautes Études en Sciences Sociales, Paris, 1991, pp. 45–62.

[299] ——, *Social choice theory and the "Centre de Mathématique Sociale": Some historical notes*, Cahiers du Centre d'Analyse et de Mathématique Sociale, 216 (2002). Forthcoming in Social Choice and Welfare.

[300] B. Monjardet and N. Caspard, *On a dependence relation in finite lattices*, Discrete Mathematics, 165–166 (1997), pp. 497–505.

[301] H. Moulin, *Axioms of Cooperative Decision Making*, no. 15 in Econometric Society Monographs, Cambridge University Press, Cambridge, 1988.

[302] ——, *Social choice*, in Handbook of Game Theory with Economic Applications: Volume 2, R. J. Aumann and S. Hart, eds., no. 11 in Handbooks in Economics, Elsevier, Amsterdam, 1994, ch. 31, pp. 1091–1125.

[303] H. M. Mulder, *The structure of median graphs*, Discrete Mathematics, 24 (1978), pp. 197–204.

[304] ——, *n-Cubes and median graphs*, Journal of Graph Theory, 4 (1980), pp. 107–110.

[305] ——, *The expansion procedure for graphs*, in Contemporary Methods in Graph Theory, R. Bodendiek, ed., BI Wissenschaftsverlag, Mannheim, 1990, pp. 459–477.

[306] H. M. Mulder and A. Schrijver, *Median graphs and Helly hypergraphs*, Discrete Mathematics, 25 (1979), pp. 41–50.

[307] Y. Murakami, *A note on the general possibility theorem of the social welfare function*, Econometrica, 29 (1961), pp. 244–246.

[308] R. B. Myerson, *Axiomatic derivation of scoring rules without the ordering assumption*, Social Choice and Welfare, 12 (1995), pp. 59–74.

[309] K. Nakamura, *The core of a simple game with ordinal preferences*, International Journal of Game Theory, 4 (1975), pp. 95–104.

[310] ——, *Necessary and sufficient conditions on the existence of a class of social choice functions*, Economic Studies Quarterly, 29 (1978), pp. 259–267.

[311] ——, *The vetoers in a simple game with ordinal preferences*, International Journal of Game Theory, 8 (1979), pp. 55–61.

[312] L. Nebeský, *Median graphs*, Commentationes Mathematicae Universitatis Carolinae, 12 (1971), pp. 317–325.

[313] G. Nelson, *Cladistic analysis and synthesis: Principles and definitions, with a historical note on Adanson's* Familles des Plantes *(1763–1764)*, Systematic Zoology, 28 (1979), pp. 1–21.

[314] D. A. Neumann, *Faithful consensus methods for n-trees*, Mathematical Biosciences, 63 (1983), pp. 271–287.

[315] D. A. Neumann and V. T. Norton, Jr., *On lattice consensus methods*, Journal of Classification, 3 (1986), pp. 225–255.

[316] M. P. NG AND N. C. WORMALD, *Reconstruction of rooted trees from subtrees*, Discrete Applied Mathematics, 69 (1996), pp. 19–31.

[317] S. NITZAN AND J. PAROUSH, *The characterization of decisive weighted majority rules*, Economics Letters, 7 (1981), pp. 119–124.

[318] S. NITZAN AND A. RUBINSTEIN, *A further characterization of Borda ranking method*, Public Choice, 36 (1981), pp. 153–158.

[319] R. D. M. PAGE, *Comments on component-compatibility in historical biogeography*, Cladistics: The International Journal of the Willi Hennig Society, 5 (1989), pp. 167–182.

[320] R. D. M. PAGE AND E. C. HOLMES, *Molecular Evolution: A Phylogenetic Approach*, Blackwell Science, Oxford, 1998.

[321] V. PARETO, *Cours d'Économie Politique*, Rouge, Lausanne, 1896. Two volumes in one.

[322] P. K. PATTANAIK, *Some paradoxes of preference aggregation*, in Perspectives on Public Choice: A Handbook, D. C. Mueller, ed., Cambridge University Press, Cambridge, 1997, ch. 10, pp. 201–225.

[323] ——, *Positional rules of collective decision-making*, in Arrow et al. [20], ch. 7, pp. 361–394.

[324] E. A. PAZNER AND E. WESLEY, *Stability of social choices in infinitely large societies*, Journal of Economic Theory, 14 (1977), pp. 252–262.

[325] I. PEARS, *An Instance of the Fingerpost*, Vintage, London, 1998.

[326] B. PELEG, *Representations of simple games by social choice functions*, International Journal of Game Theory, 7 (1978), pp. 81–94.

[327] ——, *Game-theoretic analysis of voting in committees*, in Arrow et al. [20], ch. 8, pp. 395–423.

[328] C. A. PHILLIPS AND T. J. WARNOW, *The asymmetric median tree – A new model for building consensus trees*, Discrete Applied Mathematics, 71 (1996), pp. 311–335.

[329] D. PISANI, *Comparing and Combining Trees and Data in Phylogenetic Analysis*, PhD thesis, University of Bristol, Department of Earth Sciences, June 2002. ix+313 pp.

[330] D. PISANI AND M. WILKINSON, *Matrix representation with parsimony, taxonomic congruence, and total evidence*, Systematic Biology, 51 (2002), pp. 151–155.

[331] D. PISANI, A. M. YATES, M. C. LANGER, AND M. J. BENTON, *A genus-level supertree of the Dinosauria*, Proceedings of the Royal Society of London. Series B, 269 (2002), pp. 915–921.

[332] C. R. PLOTT, *A notion of equilibrium and its possibility under majority rule*, American Economic Review, 57 (1967), pp. 787–806.

[333] ——, *Axiomatic social choice theory: An overview and interpretation*, American Journal of Political Science, 20 (1976), pp. 511–596.

[334] R. C. POWERS, *Intersection rules for consensus n-trees*, Applied Mathematics Letters, 8 (1995), pp. 51–55.

[335] ——, *Arrow's theorem for closed weak hierarchies*, Discrete Applied Mathematics, 66 (1996), pp. 271–278.

[336] ——, *Consensus n-trees and removal independence*, Journal of the Korean Mathematical Society, 37 (2000), pp. 473–490.

[337] ——, *Nondictatorially independent pairs and Pareto*, Social Choice and Welfare, 18 (2001), pp. 817–822.

[338] ——, *Consensus n-trees, weak independence, and veto power*, in Janowitz et al. [209], pp. 47–54.

[339] ——, *Medians and majorities in semimodular posets*, Discrete Applied Mathematics, 127 (2003), pp. 325–336.

[340] T. M. PRZYTYCKA, *Sparse dynamic programming for maximum agreement subtree problem*, in Mirkin et al. [286], pp. 249–264.

[341] A. PURVIS, *A modification to Baum and Ragan's method for combining phylogenetic trees*, Systematic Biology, 44 (1995), pp. 251–255.

[342] M. A. RAGAN, *Phylogenetic inference based on matrix representation of trees*, Molecular Phylogenetics and Evolution, 1 (1992), pp. 53–58.

[343] P. RAY, *Independence of irrelevant alternatives*, Econometrica, 41 (1973), pp. 987–991.

[344] M. REGENWETTER, A. A. J. MARLEY, AND B. GROFMAN, *A general concept of majority rule*, Mathematical Social Sciences, 43 (2002), pp. 405–428.

[345] J. T. RICHELSON, *A characterization result for the plurality rule*, Journal of Economic Theory, 19 (1978), pp. 548–550.

[346] W. H. RIKER, *Voting and the summation of preferences: An interpretive bibliographical review of selected developments during the last decade*, American Political Science Review, 55 (1961), pp. 900–911.

[347] F. S. ROBERTS, *Characterizations of the plurality function*, Mathematical Social Sciences, 21 (1991), pp. 101–127.

[348] ——, *On the indicator function of the plurality function*, Mathematical Social Sciences, 22 (1991), pp. 163–174.

[349] A. G. RODRIGO, *On combining cladograms*, Taxon, 45 (1996), pp. 267–274.

[350] F. Ronquist, *Matrix representation of trees, redundancy, and weighting*, Systematic Biology, 45 (1996), pp. 247–253.

[351] D. E. Rosen, *Vicariant patterns and historical explanation in biogeography*, Systematic Zoology, 27 (1978), pp. 159–188.

[352] K. A. Ross and C. R. B. Wright, *Discrete Mathematics*, Prentice-Hall, Upper Saddle River, New Jersey, fifth ed., 2003.

[353] A. Rubinstein and P. C. Fishburn, *Algebraic aggregation theory*, Journal of Economic Theory, 38 (1986), pp. 63–77.

[354] D. G. Saari, *A dictionary for voting paradoxes*, Journal of Economic Theory, 48 (1989), pp. 443–475.

[355] ——, *The Borda dictionary*, Social Choice and Welfare, 7 (1990), pp. 279–317.

[356] ——, *Consistency of decision processes*, Annals of Operations Research, 23 (1990), pp. 103–137.

[357] ——, *Calculus and extensions of Arrow's theorem*, Journal of Mathematical Economics, 20 (1991), pp. 271–306.

[358] ——, *Geometry of Voting*, no. 3 in Studies in Economic Theory, Springer-Verlag, Berlin, 1994.

[359] ——, *Informational geometry of social choice*, Social Choice and Welfare, 14 (1997), pp. 211–232. Reprinted in Heal [203, pp. 65–86].

[360] ——, *Mathematical structure of voting paradoxes: I. Pairwise vote*, Economic Theory, 15 (2000), pp. 1–53.

[361] ——, *Mathematical structure of voting paradoxes: II. Positional voting*, Economic Theory, 15 (2000), pp. 55–102.

[362] ——, *Geometry of voting*, in Arrow et al. [21], ch. 26. Forthcoming.

[363] N. Salamin, T. R. Hodkinson, and V. Savolainen, *Building supertrees: An empirical assessment using the grass family (Poaceae)*, Systematic Biology, 51 (2002), pp. 136–150.

[364] M. J. Sanderson, A. Purvis, and C. Henze, *Phylogenetic supertrees: Assembling the trees of life*, Trends in Ecology & Evolution, 13 (1998), pp. 105–109.

[365] R. Saposnik, *Social choice with continuous expression of individual preferences*, Econometrica, 43 (1975), pp. 683–690.

[366] M. A. Satterthwaite, *Strategy-proofness and Arrow's conditions: Existence and correspondence theorems for voting procedures and social welfare functions*, Journal of Economic Theory, 10 (1975), pp. 187–217.

[367] F. SCHICK, *Arrow's proof and the logic of preference*, Philosophy of Science, 36 (1969), pp. 127–144.

[368] N. SCHMITZ, *A further note on Arrow's impossibility theorem*, Journal of Mathematical Economics, 4 (1977), pp. 189–196.

[369] C. SEMPLE AND M. A. STEEL, *A supertree method for rooted trees*, Discrete Applied Mathematics, 105 (2000), pp. 147–158.

[370] A. K. SEN, *A possibility theorem on majority decisions*, Econometrica, 34 (1966), pp. 491–499.

[371] ——, *Quasi-transitivity, rational choice and collective decisions*, Review of Economic Studies, 36 (1969), pp. 381–393.

[372] ——, *Collective Choice and Social Welfare*, Holden-Day, San Francisco, 1970. Reprinted by North-Holland (Amsterdam) in 1984.

[373] ——, *The impossibility of a paretian liberal*, Journal of Political Economy, 78 (1970), pp. 152–157.

[374] ——, *Information and invariance in normative choice*, in Social Choice and Public Decision Making: Essays in Honor of Kenneth J. Arrow, Volume 1, W. P. Heller, R. M. Starr, and D. A. Starrett, eds., Cambridge University Press, Cambridge, 1986, ch. 2, pp. 29–55.

[375] ——, *Social choice theory*, in Handbook of Mathematical Economics, Volume 3, K. J. Arrow and M. D. Intriligator, eds., no. 1 in Handbooks in Economics, North-Holland, Amsterdam, 1986, ch. 22, pp. 1073–1181.

[376] K. T. SHAO, *Consensus Methods in Numerical Taxonomy*, PhD thesis, State University of New York at Stony Brook, Department of Ecology and Evolution, 1983. xix+290 pp.

[377] L. S. SHAPLEY, *Simple games: An outline of the descriptive theory*, Behavioral Science, 7 (1962), pp. 59–66.

[378] M. SHOLANDER, *Medians, lattices, and trees*, Proceedings of the American Mathematical Society, 5 (1954), pp. 808–812.

[379] L. A. SHOLOMOV, *Explicit form of neutral social decision rules for basic rationality conditions*, Mathematical Social Sciences, 39 (2000), pp. 81–107.

[380] R. SIBSON, *A model for taxonomy. II*, Mathematical Biosciences, 6 (1970), pp. 405–430.

[381] J. SLACK, *What is an explanation?*, Science, 297 (2002), p. 1813.

[382] J. H. SMITH, *Aggregation of preferences with variable electorate*, Econometrica, 41 (1973), pp. 1027–1041.

[383] R. R. Sokal and F. J. Rohlf, *Taxonomic congruence in the Leptopodomorpha re-examined*, Systematic Zoology, 30 (1981), pp. 309–325.

[384] P. S. Soltis and D. E. Soltis, *Molecular systematics: Assembling and using the Tree of Life*, Taxon, 50 (2001), pp. 663–677.

[385] M. A. Steel, *The complexity of reconstructing trees from qualitative characters and subtrees*, Journal of Classification, 9 (1992), pp. 91–116.

[386] M. A. Steel, A. W. M. Dress, and S. Böcker, *Simple but fundamental limitations on supertree and consensus tree methods*, Systematic Biology, 49 (2000), pp. 363–368.

[387] M. A. Steel and T. Warnow, *Kaikoura tree theorems: Computing the maximum agreement subtree*, Information Processing Letters, 48 (1993), pp. 77–82.

[388] E. Stensholt, *Circle pictograms for vote vectors*, SIAM Review, 38 (1996), pp. 96–119.

[389] G. A. Stephen, *String Searching Algorithms*, no. 3 in Lecture Notes Series on Computing, World Scientific, Singapore, 1994.

[390] R. Stinebrickner, *An extension of intersection methods from trees to dendrograms*, Systematic Zoology, 33 (1984), pp. 381–386.

[391] ——, *s-Consensus trees and indices*, Bulletin of Mathematical Biology, 46 (1984), pp. 923–935.

[392] ——, *s-Consensus index method: An additional axiom*, Journal of Classification, 3 (1986), pp. 319–327.

[393] P. D. Straffin, Jr., *Majority rule and general decision rules*, Theory and Decision, 8 (1977), pp. 351–360.

[394] K. Strimmer and A. von Haeseler, *Quartet puzzling: A quartet maximum-likelihood method for reconstructing tree topologies*, Molecular Biology and Evolution, 13 (1996), pp. 964–969.

[395] P. C. Suppes, *Introduction to Logic*, Van Nostrand, Princeton, New Jersey, 1957. Reprinted in 1964.

[396] ——, *Axiomatic Set Theory*, Van Nostrand, Princeton, New Jersey, 1960. Reprinted by Dover Publications (New York) in 1972.

[397] K. Suzumura, *Rational Choice, Collective Decisions, and Social Welfare*, Cambridge University Press, Cambridge, 1983.

[398] ——, *Introduction*, in Arrow et al. [20], pp. 1–32.

[399] D. L. Swofford, *When are phylogeny estimates from molecular and morphological data incongruent?*, in Phylogenetic Analysis of DNA Sequences, M. M. Miyamoto and J. Cracraft, eds., Oxford University Press, New York, 1991, ch. 14, pp. 295–333.

[400] D. L. Swofford, G. J. Olsen, P. J. Waddell, and D. M. Hillis, *Phylogenetic inference*, in Molecular Systematics, D. M. Hillis, C. Moritz, and B. K. Mable, eds., Sinauer Associates, Sunderland, Massachusetts, second ed., 1996, ch. 11, pp. 407–514.

[401] A. S. Tanguiane, *Arrow's paradox and mathematical theory of democracy*, Social Choice and Welfare, 11 (1994), pp. 1–82.

[402] J. L. Thorley and R. D. M. Page, *RadCon: Phylogenetic tree comparison and consensus*, Bioinformatics, 16 (2000), pp. 486–487.

[403] J. L. Thorley and M. Wilkinson, *A view of supertree methods*, in Janowitz et al. [209], pp. 185–193.

[404] G. Tullock, ed., *Towards a Science of Politics: Essays in Honor of Duncan Black*, Public Choice Center, Virginia Polytechnic Institute and State University, Blacksburg, Virginia, 1981.

[405] W. Vach, *Preserving consensus hierarchies*, Journal of Classification, 11 (1994), pp. 59–77.

[406] P. Vincke, *Arrow's theorem is not a surprising result*, European Journal of Operational Research, 10 (1982), pp. 22–25.

[407] R. V. Vohra, *An axiomatic characterization of some locations in trees*, European Journal of Operational Research, 90 (1996), pp. 78–84.

[408] M. S. Waterman, *Consensus patterns in sequences*, in Mathematical Methods for DNA Sequences [409], ch. 4, pp. 93–115.

[409] ——, ed., *Mathematical Methods for DNA Sequences*, CRC Press, Boca Raton, Florida, 1989.

[410] ——, *Introduction to Computational Biology: Maps, Sequences and Genomes*, Chapman and Hall, London, 1995.

[411] M. S. Waterman, R. Arratia, and D. J. Galas, *Pattern recognition in several sequences: Consensus and alignment*, Bulletin of Mathematical Biology, 46 (1984), pp. 515–527.

[412] J. S. Weber, *An elementary proof of the conditions for a generalized Condorcet paradox*, Public Choice, 77 (1993), pp. 415–419.

[413] M. Wilkinson, *Common cladistic information and its consensus representation: Reduced Adams and reduced cladistic consensus trees and profiles*, Systematic Biology, 43 (1994), pp. 343–368.

[414] ——, *More on reduced consensus methods*, Systematic Biology, 44 (1995), pp. 435–439.

[415] ——, *Majority-rule reduced consensus trees and their use in bootstrapping*, Molecular Biology and Evolution, 13 (1996), pp. 437–444.

[416] M. WILKINSON AND J. L. THORLEY, *Reduced supertrees*, Trends in Ecology & Evolution, 13 (1998), p. 283.

[417] ——, *Reduced consensus*, in Janowitz et al. [209], pp. 195–203.

[418] M. WILKINSON, J. L. THORLEY, D. PISANI, F. J. LAPOINTE, AND J. O. MCINERNEY, *Some desiderata for meta-analytical supertrees*, in Bininda-Emonds [68]. Forthcoming.

[419] R. B. WILSON, *The game-theoretic structure of Arrow's general possibility theorem*, Journal of Economic Theory, 5 (1972), pp. 14–20.

[420] ——, *Social choice theory without the Pareto principle*, Journal of Economic Theory, 5 (1972), pp. 478–486.

[421] ——, *On the theory of aggregation*, Journal of Economic Theory, 10 (1975), pp. 89–99.

[422] H. P. YOUNG, *An axiomatization of Borda's rule*, Journal of Economic Theory, 9 (1974), pp. 43–52.

[423] ——, *A note on preference aggregation*, Econometrica, 42 (1974), pp. 1129–1131.

[424] ——, *Social choice scoring functions*, SIAM Journal on Applied Mathematics, 28 (1975), pp. 824–838.

[425] ——, *Optimal ranking and choice from pairwise comparisons*, in Information Pooling and Group Decision Making: Proceedings of the Second University of California, Irvine, Conference on Political Economy, B. Grofman and G. Owen, eds., no. 2 in Decision Research, JAI Press, Greenwich, Connecticut, 1986, pp. 113–122.

[426] ——, *Condorcet's theory of voting*, American Political Science Review, 82 (1988), pp. 1231–1244.

[427] H. P. YOUNG AND A. LEVENGLICK, *A consistent extension of Condorcet's election principle*, SIAM Journal on Applied Mathematics, 35 (1978), pp. 285–300.

[428] Y. ZHOU, *A note on continuous social choice*, Social Choice and Welfare, 14 (1997), pp. 245–248. Reprinted in Heal [203, pp. 99–102].

Publication Year of References

Year	Frequency
≤1950	13
1951–1960	15
1961–1970	30
1971–1980	87
1981–1990	89
1991–2000	140
2001–2004	54
	428

Index